T0331041

Implementing Program Management

Templates and Forms Aligned with the
Standard for Program Management –
Third Edition (2013)
and Other Best Practices

Best Practices and Advances in Program Management Series

Series Editor
Ginger Levin

Construction Program Management
Joseph Delaney

Applying Guiding Principles of Effective Program Delivery
Kerry R. Wills

Program Management: A Life Cycle Approach
Ginger Levin

*Implementing Program Management: Templates and Forms Aligned
with the Standard for Program Management,
Third Edition* (2013) *and Other Best Practices*
Ginger Levin and Allen R. Green

Program Governance
Muhammad Ehsan Khan

*Successful Program Management:
Complexity Theory, Communication, and Leadership*
Wanda Curlee and Robert Lee Gordon

Sustainable Program Management
Gregory T. Haugan

The Essential Program Management Office
Gary Hamilton

*Leading Virtual Project Teams: Adapting Leadership Theories
and Communications Techniques to 21st Century Organizations*
Margaret R. Lee

From Projects to Programs: A Project Manager's Journey
Samir Penkar

Implementing Program Management

Templates and Forms Aligned with the
Standard for Program Management –
Third Edition (2013)
and Other Best Practices

Ginger Levin, PMP, PgMP
Allen R. Green, PMP, PgMP

CRC Press
Taylor & Francis Group
Boca Raton London New York

CRC Press is an imprint of the
Taylor & Francis Group, an **informa** business
AN AUERBACH BOOK

Parts of *A Guide to the Project Management Body of Knowledge*, 2013, are reprinted with permission of the Project Management Institute, Inc., Four Campus Boulevard, Newtown Square, Pennsylvania 19073-3299 U.S.A., a worldwide organization advancing the state of the art in project management.

"PgMP" is a certification mark of the Project Management Institute, Inc., which is registered in the United States and other nations.

"PMBOK" is a trademark of the Project Management Institute, Inc., which is registered in the United States and other nations.

"PMI" is a service and trademark of the Project Management Institute, Inc., which is registered in the United States and other nations.

"PMP" is a certification mark of the Project Management Institute, Inc., which is registered in the United States and other nations.

CRC Press
Taylor & Francis Group
6000 Broken Sound Parkway NW, Suite 300
Boca Raton, FL 33487-2742

© 2014 by Taylor & Francis Group, LLC
CRC Press is an imprint of Taylor & Francis Group, an Informa business

No claim to original U.S. Government works

ISBN-13: 978-1-4665-9771-6 (hbk)

Library of Congress Cataloging-in-Publication Data

Levin, Ginger.
    Implementing program management : templates and forms aligned with the standard for program management, and other best practices / Ginger Levin, Allen R. Green. -- Third edition.
      pages cm. -- (Best practices and advances in program management series)
    Includes bibliographical references and index.
    ISBN 978-1-4665-9771-6 (hardcover : alk. paper)
    1. Project management. I. Green, Allen R. II. Title.

HD69.P75L483 2013
658.4'04--dc23
                                            2013017596

Visit the Taylor & Francis Web site at
http://www.taylorandfrancis.com

and the CRC Press Web site at
http://www.crcpress.com

Dedication

To my husband, Morris, for his continuing support and love.

—Ginger Levin

To my wife, Cindy, who somehow manages to love
me back even when I work too much.

—Allen Green

Contents

Preface

Let our advance worrying become advance thinking and planning.

—**Winston Churchill**

Many of the most successful organizations worldwide have specific program management templates that they treat as carefully guarded trade secrets and consider critical to their consistent success. At this writing, it is difficult to find examples of many of the documents referenced in the Project Management Institute's (PMI®) *The Standard for Program Management*—Third Edition (2013), released December 31, 2012. When we do find roughly comparable documents here and there, they overlap or are intended for specific situations or even atypical definitions of project or program management. What is a practitioner supposed to do? What is supposed to be in the various documents? How do they complement rather than overlap each other?

We first wrote a comparable book in 2010, after the Second Edition of the Standard was released. In December 2008. Since there were only a few program management books available at that time, our goal was for it to be an attempt to lift the shroud of obscurity surrounding such documents. This goal remains today.

While there are more program management books available in 2013, the Third Edition of the PMI Standard provides less detail and a different focus than that of the Second Edition. We have updated templates and forms in this book to complement what is included in the Third Edition as well as to include others that we feel are best practices to use in managing programs.

The templates and forms herein are examples to consider for making your next programmatic adventure a little less reminiscent of a sequel to Raiders of the Lost Ark. We want this book to be a liberating breath of fresh air that breaks the writer's block so commonly encountered in the program management process. We feel it will also be extremely valuable to students of program management, who forever request "go-by" templates for their work, and rightfully so.

The professions (and we dare say arts) of project and program management are moving toward maturity and standards. PMI's *A Guide to the Project Management Body of Knowledge*—Fifth Edition (2013) and *The Standard for Program Management*—Third Edition (2013) are already standards of the American National Standards Institute (ANSI). Further, the International Organization for Standardization (ISO) effort has been completed as a global standard for project management knowledge and processes. PMI had an active role in that effort. Various organizations worldwide will likely position themselves as consistent with the ISO standard. Program and portfolio management are not currently on the ISO radar screen, but their standardization would seem a natural next step.

In this book the best practices we have included must be adapted in multiple ways when used on any specific program:

1. For the given industry and industry segment
2. For the culture and practices of a given organization
3. For the preferences of the program manager and program team
4. For the target program and its environment, considering its purpose and scale

That is a lot of adaptation! Program documents are a little like a covered dish supper at Grandma's house in that each planner has unique tastes.

As just one illustration, consider the Defense Acquisition Guidebook published by the U.S. Department of Defense. Accommodating multi-billion-dollar programs and having complex stakeholder relationships, many consider that program management discipline to be completely different. In fact, it has much in common with *The Standard for Program Management*—Third Edition (2013) at its core, and several plans herein are valuable elaborations of its prescribed documents.

Many program practitioners may already have commercial off-the-shelf or homegrown systems in place that implement some of the concepts we illustrate here. We hope this book spurs ideas for improvement of those systems as well.

So it is that your documents will likely look quite different. Bravo!

Ginger Levin, PMP®, PgMP®
Lighthouse Point, Florida

Allen R. Green, PMP®, PgMP®
Huntsville, Alabama

Acknowledgments

We want to acknowledge the efforts of our publisher, CRC Press, and especially that of John Wyzalek, Randy Burling, Jessica Vakili, and the entire CRC team who worked tirelessly to publish this book so it would be available according to the release of the new PgMP® exam from PMI®.

File Download

Templates and forms are available for download at http: //www.ittoday.info/pgmp/ImplementingFiles.html

To download the files, use the six-character code above the barcode on the book's back cover.

Introduction

Discipline is the bridge between goals and accomplishment.

—Jim Rohn

This book is a workbook for successful program management. Success in program management requires discipline—the kind of discipline that results in repeatability of success. It requires appropriately complete plans, well-run meetings with meaningful agendas, accurate record keeping, and general adherence to global best practices. Toward those ends, this book is a compendium of plans, forms, agendas, registers, and procedures.

This work is specifically designed to be compatible with and in a supplementary to the Project Management Institute's (PMI®) *The Standard for Program Management*—Third Edition (2013). Starting out to produce artifact prototypes for all the documents mentioned therein, we found that the task proved a bit more elusive than expected. Some documents are explicitly enumerated in *The Standard for Program Management*—Third Edition (2013), and others are more implicit—so we have made judgments as to what to include. We have included almost all of the identifiable documents referenced in *The Standard for Program Management*—Third Edition (2013). We have also kept some useful ones from prior editions that are now omitted. We also added some of our own. In a couple of instances, *The Standard for Program Management*—Third Edition (2013) defers to PMI's *A Guide to the Project Management Body of Knowledge*—Fifth Edition (PMBOK®), and so do we.

This work is not intended to be a complete guide to program management. Nor will it be entirely useful as a supplementary study for PMI's Program Management Professional (PgMP®) credential examination. You will need *The Standard for Program Management*—Third Edition (2013), the Examination Content Outline or ECO (2011), and some of the other available texts and articles on program management. We make no assumptions about your level of experience in program management but accept the fact that you may or may not be new to the discipline.

Our focus is on three heretofore neglected agendas: (1) helping you apply *The Standard for Program Management*—Third Edition (2013) and the other best practice guidelines included in this book in the successful execution of your program on a daily basis, (2) meeting the needs of undergraduate and graduate students at the university level and those in short courses in program management who may require guidance in their initiating, planning, executing, monitoring and controlling, and closing exercises, and (3) serving as a reference book for PgMP aspirants with its organization according to the program management domains as specified in the ECO.

About Program Management

Although projects have been under way since the time of the pyramids, and project management emerged as a profession in the late 1950s and early 1960s, we now work in a project-oriented society. However, rather than having only one or two large projects under way, organizations have hundreds of projects of various sizes and complexities. Organizations began to view projects as assets and started practicing management by projects.

The emphasis, though, now has shifted to program management.

If projects, and other ongoing work, can be grouped into a program because they share common attributes, the organization can then realize more benefits than if the projects were managed in a stand-alone fashion, resources can be coordinated across projects, customer relationship management is improved, as well as supply chain management, and greater effectiveness and efficiencies can result.

Recognizing the increasing use of programs, in 2006, PMI issued a standard on program management, which was updated at the end of 2008, and again updated at the end of 2012, and published in 2013. This standard describes best practices in program management and contains a common set of terms to best communicate globally among the various program stakeholders. These best practices are of enormous value, as they show the importance of key documents to initiate a program, plan it, execute and monitor and control it, and close it. We now are working in a society in which program management is the norm and not the exception and in which programs and projects are considered organizational assets. We are managing by both programs and projects.

Let's get to work!

Acronyms

CRM	Customer relationship management
DoD	U.S. Department of Defense
ECO	Examination Content Outline
KM	Knowledge management
MIL-HDBK	Military Standard
PWBS	Program work breakdown structure
PMBOK®	Project Management Body of Knowledge
PMI®	Project Management Institute
PgMP®	Program Management Professional
PM	Program manager
PMP®	Project Management Professional
RAM	Responsibility assignment matrix
RACI	Responsible, accountable, consult, inform
SWOT	Strengths, weaknesses, opportunities, and threats

About the Authors

Ginger Levin, DPA, is a senior consultant and educator in project management. Her specialty areas are portfolio management, program management, the project management office, metrics, and maturity assessments. She is certified as a PMP®, PgMP®, and *OPM3* (Organizational Project Management Maturity Model) Professional.

In addition, Dr. Levin is an adjunct professor in project management for the University of Wisconsin–Platteville in its Master of Science degree program and for SKEMA University, Lille, France, and RMIT in Melbourne, Australia, in their doctoral project management programs.

In consulting, she has served as program or project manager in numerous efforts for Fortune 500 and public sector clients, including Cargill, Medtronic, Abbott Vascular, MARTA, UPS, Citibank, the Food and Drug Administration, General Electric, SAP, EADS, John Deere, Schreiber Foods, TRW, New York City Transit Authority, and the U.S. Department of Agriculture. Prior to her work in consulting, she held positions of increasing responsibility with the U.S. government, including the Federal Aviation Administration, the Office of Personnel Management, and the General Accounting Office.

She is the editor of *Program Management: A Life Cycle Approach*, the author of *Interpersonal Skills for Portfolio, Program, and Project Managers*, co-author of *Program Management Complexity: A Comptency Model, Implementing Program Management Templates and Forms Aligned with The Standard for Program Management*—Second Edition (2008) [with Allen Green], *Project Portfolio Management, Metrics for Project Management, Achieving Project Management Success with Virtual Teams, Advanced Project Management Office: A Comprehensive Look at Function and Implementation, People Skills for Project Managers, Essential People Skills for Project Managers, The Business Development Capability Maturity Model, PMP® Challenge!, PMP® Study Guide, PgMP® Study Guide,* and *PgMP® Challenge!*.

Dr. Levin received her doctorate in information systems technology and public administration from The George Washington University and received the Outstanding Dissertation Award for her research on large organizations. She also received her master of science in business administration from The George Washington University and her bachelor of business administration from Wake Forest University.

Allen R. Green, MS, PMP®, PgMP®, of Science Applications International Corporation–Frederick has broad and extensive leadership experience in the commercial, military, health care, National Aeronautics and Space Administration, civil service, civic, local government, and academic sectors. Throughout his career, he has specialized in initial build-out of the various functional areas of new programs.

He led software development projects at Capability Maturity Model Integrated (CMMI) levels II through V for twelve years, including the first delivered applications for the U.S. Army's Future Combat System. His additional training experience included tours as president and management development vice president of the Alabama Junior Chamber of Commerce, metropolitan vice president of the U.S. Junior Chamber of Commerce, and president of the Alabama Junior Chamber of Commerce Foundation, with responsibility for training of project planners. As vice chairman of the Madison County Emergency Communications (E-911) Board, he chaired the effort to establish a unified state-of-the-art centralized call center in Madison County, Alabama, one of the first nationally. In the commercial sector, he managed the development program for the first shrink-wrapped electronic software distribution suite for industry standard microcomputers.

He is a past president of Project Management Institute's North Alabama Chapter. As the chapter's professional development vice president he organized and participated in the training of hundreds of Project Management Professional (PMP®) and Program Management Professional (PgMP®) candidates. He has taught software engineering and program management under contract to several universities in the United States and the Middle East.

He received PMI's PgMP® credential in August 2008 and holds Bachelor of Electrical Engineering and Master of Science in electrical engineering degrees from Auburn University.

Chapter 1

Introduction

Discipline is the bridge between goals and accomplishment.

—Jim Rohn

This book is a workbook for successful program management. Success is program management requires discipline—the kind of discipline that results in repeatability of success. It requires appropriately complete plans, well-run meetings with meaningful agendas, accurate record keeping, and general adherence to global best practices. Toward those ends, this book is a compendium of plans, forms, agendas, registers, and procedures.

This work is specifically designed to be compatible with and is a supplementary to the Project Management Institute's (PMI®) *Standard for Program Management—* Third Edition (2013). Starting out to produce artifact prototypes for all the documents mentioned therein, we found that the task proved a bit more elusive than expected. Some documents are explicitly enumerated in the *Standard for Program Management*—Third Edition (2013), and others are more implicit—so we have made judgments as to what to include. We have included almost all of the identifiable documents referenced in the *Standard for Program Management*—Third Edition (2013). We have also kept some useful ones from prior editions that are now omitted. We also added some of our own. In a couple of instances, the *Standard for Program Management*—Third Edition (2013) defers to PMI's *A Guide to the Project Management Body of Knowledge (PMBOK Guide)*—Fifth Edition (2013) (PMBOK Guide, Fifth Edition—2013), and so do we.

This work is not intended to be a complete guide to program management. Nor will it be entirely useful as a supplementary study for PMI's Program Management

Professional (PgMP®) credential examination. You will need the *Standard for Program Management*—The Edition (2013), the *Examination Content Outline* or ECO (2011), and some of the other available texts and articles on program management. We make no assumptions about your level of experience in program management but accept the fact that you may or may not be new to the discipline.

Our focus is on three heretofore neglected agendas: (1) helping you apply the *Standard for Program Management*—Third Edition (2013) and the other best practice guidelines included in this book in the successful execution of your program on a daily basis, (2) meeting the needs of undergraduate and graduate students at the university level and those in short courses in program management who may require guidance in their initiating, planning, executing, monitoring and controlling, and closing exercises, and (3) serving as a reference book for PgMP aspirants with its organization according to the program management domains as specified in the ECO.

About Program Management

Although projects have been under way since the time of the pyramids, and project management emerged as a profession in the late 1950s and early 1960s, we now work in a project-oriented society. However, rather than having only one or two large projects under way, organizations have hundreds of projects of various sizes and complexities. Organizations began to view projects as assets and started practicing management by projects.

The emphasis, though, now has shifted to program management.

If projects, and other ongoing work, can be grouped into a program because they share common attributes, the organization can then realize more benefits than if the projects were managed in a stand-alone fashion, resources can be coordinated across projects, customer relationship management is improved, as well as supply chain management, and greater effectiveness and efficiencies can result.

Recognizing the increasing use of programs, in 2006, PMI issued a standard on program management, which was updated at the end of 2008, and again updated at the end of 2012 and published in 2013. This standard describes best practices in program management and contains a common set of terms to best communicate globally among the various program stakeholders. These best practices are of enormous value, as they show the importance of key documents to initiate a program, plan it, execute and monitor and control it, and close it. We now are working in a society in which program management is the norm and not the exception and in which programs and projects are considered organizational assets. We are managing by both programs and projects.

Let's get to work!

Chapter 2

Getting Started

> In preparing for battle I have always found that plans are useless, but *planning is indispensable.*
>
> **—Dwight D. Eisenhower**

Planning is woefully implemented in some programs. On the other hand, in certain cultures, planning is almost an (endless) end in itself—sort of like the infamous "paralysis by analysis." Your world needs to be somewhere in between.

An important take-home message here is that, given natural and often turbulent change in a program's environment, planning in too much detail can be a counterproductive drain on your time. Planning in too little detail, however, can cause you to overlook things that need to be at least thought through in advance. One of the most important program manager skills is the ability to do just the right amount of all the right things and at the right time.

If the documents we describe are new to your organization, almost certainly the first application will be the most difficult. Major parts of your work will become what the Project Management Institute (PMI®) refers to as organizational process assets. So if you are in that situation, you will be a pioneer. As they say about pioneers, they are the ones with the arrows in their back. That is sort of like us perhaps, in even daring such a work as this book.

Why So Many Plans?

The seeming paradox that leads off this chapter actually makes sense! How?

During the often furious pace of a program's life time, indeed things do change. And unmanaged change is your enemy, while considered and managed change is your friend. When it occurs, your prior planning work has provided you with an understanding of the program and proves invaluable in defining and implementing the necessary changes.

In numerous templates that follow, in each phase of the Program Life Cycle domain, we begin in each one with a brief statement of its value, followed by a detailed description of its contents, and then a template, ready to tailor and use.

Since we have organized this book based on the five domains in the Examination Content Outline (ECO), we have included plans that we recommend in benefits, stakeholders, and governance in their specific chapters and other templates in this book so there are even more templates to consider!

First Steps

As in all endeavors, it is best to take advantage of preexisting assets. As pointed out in the preface, your organization is unique. In some cases, you will need only to reference existing standard documents for your organization. In other cases, you may need to prepare extensive documents using ours as a starting point. And in yet other cases, you might only need to include a paragraph as part of another document, indicating how due consideration has been given to the matter at hand.

First, consult with your program management office, if available, or senior management to get electronic access to any templates, forms, and standards for program, quality, financial management, etc., and other plans that you can adapt for your program. Depending on the program and project management maturity of your organization, these items may be fairly easily obtained. If all else fails, you may be able to locate suitable plans for similar programs in the past, or just begin with the templates we provide. In some cases, you will be able to refer to external standards by reference, so long as they are appropriately archived in your program repository.

Next, where applicable, gather program-unique standards that your customer requires. Being able to refer to these items may make your writing experience considerably easier.

Chapter 3

Strategic Program Management/Alignment

Luck is what happens when preparation meets opportunity.

—Seneca

In this phase there is as yet no program—that is, a program charter does not yet exist. In fact, the purpose of this phase is to justify the program, usually through an approved the business case. A description of the benefits to be achieved, a list of preliminary identified stakeholders, and the program's mission statement are produced as well in order to convince the organization's program selection process to accept the effort and to select it as a program rather than as a series of projects.

Our goal is to ensure that our programs not only provide more benefits than if the projects and other work within them were managed separately, but also to ensure that our programs align to our organization's strategic objectives. In many cases, this alignment sounds easy but in actuality is quite difficult.

While each organization does have a portfolio representing all of the work that is under way, one may not know the contents of the portfolio, or how strategic decisions are made to determine whether or not to add a new program or project to the portfolio, unless it is a mature organization with a well-defined portfolio management process in place. Limited resources make portfolio management an imperative to ensure that the programs and projects that are selected to be part of the portfolio are the "right" ones to pursue given the organization's strategic goals and objectives. Later, portfolio management ensures that chosen portfolio programs, projects, and ongoing activities are ones that should continue to be pursued.

5

According to the *Examination Content Outline* (PMI, 2011), a high-level view of the key tasks in Strategic Program Management is as follows:

- Performing an initial assessment to ensure the program is aligned to the strategic plan, its objectives, priorities, and vision or "to be" state, and conforms to its mission or reason for importance
- Preparing a high-level roadmap with milestones
- Using this roadmap as a way to set a baseline to define the program and support its planning and execution
- Evaluating the capability of the organizational leaders to assess the program's objectives, ensure their priority, feasibility, readiness, and alignment to strategic goals
- Identifying with a variety of techniques the benefits to the organization from the program
- Estimating these benefits both in financial and non-financial ways
- Evaluating program objectives according to regulatory and legal constraints, social impacts, sustainability, and other concerns
- Obtaining approval from organizational leaders to approve the program
- Identifying and evaluating opportunities for integration of the program's activities to further align and integrate the program's benefits within and across the organization
- Evaluating strategic opportunities for change to maximize benefits for the organization flowing from the program

The Standard for Program Management—Third Edition (2013) refers to this domain as Program Strategy Alignment, focusing as well on organizational strategy and the alignment of the program to it along with environmental assessments. The program, therefore, emphasizes identifying both opportunities and benefits to accomplish the organization's strategic objectives once the program is implemented. It notes the importance of determining on the need for the program and in doing so to validate the proposed outcomes through preparation of its business case. A program roadmap is another best practice to prepare with a strategic plan for the program.

We next will take a look at the artifacts produced in the Strategic Program Management or Program Strategic Alignment domain.

Program Benefits Statement

The benefits statement for the program is a key document. Without a practical and useful statement, there is no reason to have a program or to bother preparing a business case for it. In Strategic Program Management, the high-level benefits are set forth for the program, including both financial and non-financial benefits, thus assisting the obtainment of approval for the program and helping to determine its priority in the organization's portfolio. The benefits identified then are included in the program's business case.

Program Benefits Statement Instructions

The program benefits statement includes the following:

Purpose: A brief introductory statement defining the purpose of the benefits statement, such as:

> The program benefits statement defines the benefits for pursuing a program by documenting exactly what will be accomplished.
>
> Its purpose is to address the value and organizational impact of the program. It is an iterative document prepared before the program is authorized to be part of the program and moves to the initiating phase or program definition phase of the life cycle. It also is used to define benefits that later will be expanded as the benefits realization plan is prepared. It assists in developing the business case, program charter, and the program management plan and supports overall program monitoring and controlling.

Objectives and success criteria: This section formalizes the benefits to be realized by the program, both tangible and intangible.

Assumptions: This section lists and describes the specific program assumptions associated with the program benefits and the potential impact of the assumptions should they prove to be false.

Constraints: This section lists and describes the specific program constraints associated with the program benefits that will limit the team's options.

Interdependencies between projects in the program: This section identifies interdependencies between the benefits in the various proposed and existing projects that will comprise the program, as one project may produce benefits that then are used by another project later in the program.

Changes: This section identifies the potential impact of program changes on benefit outcomes.

Approvals: This section contains the written approval of the program benefits statement by the program sponsor, the program management office, members of the Portfolio Review Board or comparable group, and other stakeholders.

Initial Program Benefits Statement Template

<Insert Program Name>
Initial Program Benefits

Program name:	
Program sponsor:	
Proposed start date:	
Proposed end date:	
Prepared by:	
Revision history:	
Business unit:	

A. PURPOSE

A brief introductory statement defining the purpose of the benefits statement, such as:

> The program benefits statement defines the benefits for pursuing the program by documenting them with available information.

B. OBJECTIVES AND SUCCESS CRITERIA

This section lists the benefits to be achieved, both tangible and intangible or intrinsic or extrinsic.

C. ASSUMPTIONS

This section lists and describes any assumptions that may impact benefit realization, and the potential impact of the benefits if they prove to be false.

D. CONSTRAINTS

This section lists and describes any constraints associated with the proposed benefits that would limit the options of the program team.

E. INTERDEPENDENCIES BETWEEN PROJECTS IN THE PROGRAM

This section identifies interdependencies between the benefits in the proposed projects to be part of the program as one project may produce benefits that are used later by another project in the program.

F. CHANGES

This section identifies potential impacts of program changes on benefit outcomes.

G. APPROVALS

This section contains the approval of the benefits statement by the program sponsor and others as required.

SIGNATURES AND DATE APPROVAL OBTAINED

Program sponsor _____

Person 1 _____

Person 2 _____

Person N _____

Initial Stakeholder Identification

To be successful, you must identify in a timely fashion all those people who can help, hurt, or perhaps even kill your program before or after its birth. You must ultimately deal with each one appropriately. Skip this step at your peril.

This document is an initial list of program stakeholders. It is actually referenced later in this book in the Stakeholder Domain and developed more completely as the stakeholder register, but we include it here because it is so important before the program is approved officially.

Initial Stakeholder Identification Instructions

The initial stakeholder identification document includes the following:

Purpose: A brief introductory statement defining the purpose of the initial stakeholder identification, such as:

Stakeholder engagement is a key theme in program management. It is necessary to identify the stakeholders who will have an interest in or involvement with the program as early as possible to build support for the program and to show with whom communication is required.

Different stakeholders will have an interest in or influence over the program at different times. When the program is proposed for consideration, stakeholder identification begins. It then is used to support the business case, as it helps to build support for the program and assists in the development of the program stakeholder engagement plan and the program communications management plan. It also assists in the development of the overall program management plan. Further, as the program's mission statement describing why the program is important is prepared, stakeholder concerns and expectations require consideration to best address the overall direction of the program.

At times, programs are established because a number of the organization's projects involve the same stakeholders. Stakeholder involvement is a prerequisite for program success.

Stakeholder definition: Different organizations will define stakeholders in different ways; a standard definition may be provided by a program management office. This section states what is meant by a stakeholder, as a clear definition is required to determine whether the relevant stakeholders are identified. Note that the definition may be a person or group with an interest in the program, involvement (passive or active) in the program, influence over the program, provides information to the program, or is otherwise affected by the program. Stakeholders may be internal or external to the program.

Stakeholder roles: This section defines the various roles of program stakeholders, noting the importance of involving those stakeholders who are expected to be more active than others throughout the program life cycle. Roles can be based on functions to be performed, such as making program decisions, having responsibility for a program package, providing funding, providing resources, providing expert advice and consultation as appropriate, defining benefits and outcomes, ensuring that customer or user requirements are met, or contributing to the outcomes. This section states the names or groups involved in each identified role, and it provides contact information as appropriate.

Phases of the program life cycle: This section shows the roles, specific stakeholders, and their expected levels of involvement in the various phases of the program life cycle. A classification system such as high, medium, or low; active versus passive; or positive versus negative may be used.

Stakeholder communication methods: This section provides initial information as to methods to communicate with each of the identified stakeholders to obtain their support for the program. It will be further elaborated in the program stakeholder engagement plan and the communications management plan. It is included to engage those key stakeholders in initial support for the program.

Approvals: This section contains the written approval of the initial stakeholder identification by the program sponsor and others as appropriate.

Initial Stakeholder Identification Template

<Insert Program Name>
Initial Stakeholder Identification

Program name:	
Program sponsor:	
Proposed start date:	
Proposed end date:	

Prepared by:	
Revision history:	
Business unit:	

A. PURPOSE

A brief introductory statement defining the purpose of initial stakeholder identification, such as:

> Stakeholder engagement is a key theme in program management. It is necessary to identify the stakeholders who will have an interest in or involvement with the program as early as possible to build support for the program, define the program mission statement, and to show with whom communication is required.

_____

_____

_____

B. STAKEHOLDER DEFINITION

This section states what is meant by a stakeholder of the program, as a clear definition is required to determine whether the relevant stakeholders are identified.

_____

_____

_____

C. STAKEHOLDER ROLES

This section defines the various roles of program stakeholders, noting the importance of involving those stakeholders who are expected to be more active than others throughout the program's life cycle. It states the names or groups involved in each identified role and provides contact information as appropriate.

_____

_____

_____

D. PHASES IN THE PROGRAM LIFE CYCLE

This section shows the role of specific stakeholders and their expected levels of involvement in the various phases of the program life cycle.

_____

_____

_____

E. STAKEHOLDER COMMUNICATIONS METHODS

This section provides initial information as to methods to communicate with each of the identified stakeholders to obtain their support for the program.

_____

_____

_____

F. APPROVALS

This section contains the approval of the initial stakeholder identification template by the program sponsor and others as required.

SIGNATURES AND DATE APPROVAL OBTAINED

Program sponsor   _____

Person 1   _____

Person 2   _____

Person N   _____

Program Business Case

The program business case is used as the key document to obtain organizational approval for the program. It describes, among other things, the program's cost-benefit justification. Quite naturally, this information is used and updated throughout the overall life cycle of the program. It is described in *The Standard for Program Management*—Third Edition (2013) as the key document to be used to charter and authorize programs. Updates to the business case are made as required and approved if the program is authorized. Further, it is reviewed regularly especially as part of the reviews by the Portfolio Review Board or comparable group to determine whether or not the program should continue to be part of the portfolio to ensure it still is contributing to the program's goals and objectives. When the program is ready to close, the business case is reviewed to assess the benefits stated in it with those that actually were realized.

In the *Standard for Program Management*—Third Edition (2013), the business case is part of organizational strategy and program alignment. It is used in:

- Program mandate preparation (3.1.1)
- Benefit identification and in preparing the benefits register (4.1.2)
- Stakeholder engagement planning (5.2)
- Program approval, endorsement, and initiation (6.2.2)

It is updated as follows and as required:

- After the program financial framework is prepared (8.2.2)
- After the program charter is prepared (8.3.1.5)

Program Business Case Instructions

Purpose: A brief introductory statement defining the purpose of the program business case, such as:

> The program business case describes the need, feasibility, and justification for the program.

Typically, this document is prepared by the program sponsor with input from other key stakeholders. In some organizations, the program manager may have been identified, and he or she also participates in its development. It may be provided by the client or the funding organization. It is prepared before the program is approved, as it is the key document used in selecting the program for approval by the organization's Portfolio Review Board or other decision-making body. Once approved, the program initiation or program definition phase begins. It then provides direction for the program and shows the value the program should deliver.

Although the business case is prepared to initiate the program, it should be revisited and updated as required on a regular basis to ensure that the program continues to meet its objectives and is delivering its expected benefits.

Program strategic objectives: This section describes the expected results from the program. It shows how these expected results complement the organization's strategic objectives. It states the vision or end state of the program, the mission or why the program is important, and the values to be used in making decisions concerning program activities. The program's objectives must be aligned with the organization's strategic plan and goals.

Program benefits: A benefit is an improvement to the running of an organization that provides utility to the organization and/or customers or other groups. Benefits may be financial or nonfinancial, intrinsic or extrinsic, tangible or intangible. This section describes the expected benefits to be delivered through the program, such as to enhance existing capabilities or develop new capabilities. It explains why it is more appropriate to manage this initiative as a program rather than as separate projects, and describes the types of projects that probably will be part of the program and the collective benefits to be realized through a program management structure. It identifies the level of investment and support required to realize the proposed benefits.

Required resources: This section provides a high-level estimate of the resources required for the program. It includes all resources, not just human resources, and shows why they are needed and when they will need to be available. It highlights the value of the program against the resources that are required for its successful implementation.

Financial analysis: This section contains the financial analysis of the program in terms of financial indices, such as the return on investment, net present value, and the payback period. It shows the costs to execute the program compared to the benefits to be realized by the program. These indicators should be tracked throughout the program. This section describes sources of program funding, trends in labor and material costs, and contract costs as appropriate.

Assumptions and constraints: Programs are based on a series of assumptions or facts that are considered to be true, real, or certain. They also include constraints or those factors that may be imposed on a program such as available funding, legislative actions, or a time-to-market deadline. This section lists assumptions and constraints applicable to the program.

Internal and external influences: Programs may be initiated for a number of reasons, and there are a wide range of internal and external influences that can affect the program or show the reason it should be undertaken. This section describes these influences and explains the program's business and operation impact. If it is an external program, this section states the market demand for the program, environmental impacts, social needs, and any barriers to entry. Time-to-market data are included in this section. In addition, these influences may affect the execution of the program. This section describes these influences and why they are significant to the program, as they often become program constraints.

Alternative analysis: Because of resource limitations and other constraints, it is necessary to recognize that competing initiatives exist in the organization as all possible projects and programs cannot be approved. This section describes the results of alternative analyses, or "what if" analyses, to show different approaches that could be considered in order to meet the program's intended objectives. Feasibility studies may be included as well in this section. It is used to show that initiating this program is the most effective way to meet these objectives and deliver the expected benefits. Strength, weakness, opportunity, and threat (SWOT) analysis may be included.

Program complexity: Programs are initiated because of the need for some type of change to implement new capabilities to be used within the organization or by customers or to enhance existing capabilities. This section describes the complexity of this change initiative to best determine whether or not it should be managed as a program and undertaken by the organization.

Approvals: This section contains the written approval of the program business case by the members of the Portfolio Review Board or similar group and any other key stakeholders as appropriate. These approvals justify the use of resources required for the program and project the value the program is to deliver to the organization.

Program Business Case Template

<Insert Program Name>
Program Business Case

Program name:	
Program sponsor:	
Proposed start date:	
Proposed end date:	
Date prepared:	
Revision history:	
Business unit:	

A. PURPOSE

A brief introductory statement defining the purpose of the business case, such as:

> The program business case describes the need, feasibility, and justification for the program.

_____

_____

_____

B. PROGRAM STRATEGIC OBJECTIVES

This section describes the expected results from the program and how they complement the organization's strategic objectives.

_____

_____

_____

C. PROGRAM BENEFITS

This section describes the expected benefits to be delivered by the program. It describes why it is more appropriate to manage this initiative as a program rather than as separate projects.

_____

_____

_____

D. REQUIRED RESOURCES

This section provides a high-level estimate of the resources required for the program. It highlights the value of the program against the resources that are required for its successful implementation.

_____

_____

_____

E. FINANCIAL ANALYSIS

This section describes the results of the financial analyses that were conducted and describes sources of program funding, trends in labor and material costs, and contract costs as appropriate.

_____

_____

_____

F. ASSUMPTIONS AND CONSTRAINTS

This section describes the assumptions and constraints that may affect the program if it is approved.

_____

_____

_____

G. INTERNAL AND EXTERNAL INFLUENCES

This section describes the internal and external influences that show why the program should be undertaken and also that may affect the program if it is approved.

_____

_____

_____

H. ALTERNATIVE ANALYSIS

This section describes the results of any alternative analyses that have been undertaken to show different approaches that could be used in order to meet the program's intended objectives. Any feasibility studies that were undertaken can be attached to this section.

_____

_____

_____

I. PROGRAM COMPLEXITY

This section describes the complexity of the initiative to determine whether or not it should be managed as a program and undertaken by the organization.

_____

_____

_____

J. APPROVALS

This section contains the approval of the program business case by the members of the Portfolio Review Board and other key stakeholders as required.

SIGNATURES AND DATE APPROVAL OBTAINED

Portfolio Review Board chairperson _____

Portfolio Review Board member 1 _____

Portfolio Review Board member 2 _____

Portfolio Review Board member 3 _____

Stakeholder 1 _____

Stakeholder 2 _____

Stakeholder N _____

Program Mandate

The program mandate, along with the business case, typically is the major document that will be used to gain approval to proceed to program initiation. It is a key input for the Portfolio Review Board or comparable group use to determine whether to charter and authorize programs. As *The Standard for Program Management—Third Edition* (2013) notes, it is used to confirm commitment of organizational resources to the program in that it defines the program's expected strategic objectives and benefits. It is used in the benefits identification process (4.1). In some organizations, the term Program Brief is used. It is issued by the Portfolio Review Board or a comparable group.

Program Mandate Instructions

The program mandate includes the following:

Purpose: A brief introductory statement defining the purpose of the program mandate, such as:

> The program mandate describes the importance for this program to be part of the organization's portfolio.

> The program mandate is typically prepared after the business case has been made for the program and before the program is initiated. The program sponsor usually prepares the program mandate, since he or she is the person who will provide funding for the program, with input from other key stakeholders. A client may suggest that the program be undertaken and may provide supporting details concerning its importance. It serves as the document that is provided to members of the organization's Portfolio Review Board or comparable group to approve the program as it defines the strategic objectives and benefits and then leads to the formal initiation process.
>
> This template provides guidance for the generic factors that should be included as part of the program mandate, recognizing that it will be tailored, since each organization will have different criteria that it uses to select and prioritize programs, projects, and ongoing operations. The program mandate should address these specific criteria used by the organization.

Program importance: This section sets the stage for the remainder of the document, as it shows the importance of the program to the organization. It provides a description of what the program is intended to achieve and why this initiative should be managed as a program rather than as a project.

Program vision: This section contains the vision for the program or its end state, providing more details as to what the program will achieve.

Program strategic objectives: This section states the strategic objectives of the program and shows how each objective relates to the organization's strategic objectives and strategic plan. It shows how these objectives then will achieve the program's vision. This section also shows how this program will interface with other programs and projects in the organization that are under way and are part of the overall strategic initiatives.

Program benefits: A benefit is an outcome of the program, such as an improvement to the overall operations of the organization or a desired product, service, or result for clients. This section describes the benefits of the program to the organization and how they will be realized through the various components of the program—projects and non-project work.

Program scope and components: This section states the scope of the program in terms of the products, services, and results that will be delivered in order to realize the benefits. It describes the identified components that will comprise the program and how they will relate to one another; often these components already are under way as individual projects. If this is the case, this section shows how these components, if managed as a program, will achieve more benefits than if they were continued as individual projects. This section also describes what is considered to be outside of the scope of the program.

Assumptions and constraints: This section states the assumptions and constraints that are expected to affect the proposed program. Assumptions are those items that are considered to be true, real, or certain without any backup documentation, while constraints represent any milestones to be imposed on the program or limits in terms of schedule, costs, and resources.

Program schedule: This section presents a high-level schedule of the program's deliverables and key milestones.

Resource requirements: This section states the resource requirements for the program and includes all resources, not just people. It also notes when specific resources will be required according to the high-level schedule.

Risks and issues: Various risks and issues will affect the program. This section presents those risks and issues that may affect the program, positively or negatively, and that have been identified at the time the program is proposed in order that the organization can determine whether the level of risk is acceptable before proceeding further.

Stakeholder considerations: Stakeholders play a key role in any program. This section presents a list of those stakeholders that are expected to influence the program, positively or negatively, and others who will have an interest in the program at certain phases in the life cycle.

Governance requirements: This section states the governance requirements for the program as it specifies the use of a Governance Board or similar group and suggests key members for this Board. It states whether it is necessary to modify existing governance procedures used in the organization for the program and shows how the Board will meet to oversee the program according to the schedule. This section also states the governance processes at the program level that are proposed for oversight of the various components. It describes reporting requirements to the Governance Board from the program manager and to the program manager from component managers.

Approvals: This section contains the written approval of the program mandate by the members of the Portfolio Review Board or similar group and any other key stakeholders as appropriate.

Program Mandate Template

<Insert Program Name>
Program Mandate

Program name:	
Program sponsor:	
Proposed start date:	
Proposed end date:	
Date prepared:	
Revision history:	
Business unit:	

A. PURPOSE

A brief introductory statement defining the purpose of the program mandate, such as:

> The program mandate describes the importance for this program to be part of the organization's portfolio.

_____

_____

_____

B. PROGRAM IMPORTANCE

This section provides a description of the program and why it is important to the organization. It also states why the initiative should be managed as a program rather than as individual projects.

_____

_____

_____

C. PROGRAM VISION

This section contains the vision for the program or its end state.

_____

_____

_____

D. PROGRAM STRATEGIC OBJECTIVES

This section states the strategic objectives of the program and shows how each objective relates to the organization's strategic objectives and strategic plan. It shows how these objectives will achieve the program's vision and describes how this proposed program relates to other programs and projects under way in the organization.

_____

_____

_____

E. PROGRAM BENEFITS

This section describes the benefits of the program to the organization and how they will be realized through the various program components.

_____

_____

_____

F. PROGRAM SCOPE AND COMPONENTS

This section states the scope of the program in terms of the products, services, and results to be delivered to realize benefits. It also describes what is considered to be outside of the scope of the program.

_____

_____

_____

G. ASSUMPTIONS AND CONSTRAINTS

This section describes the assumptions and constraints that are expected to affect the proposed program.

_____

_____

_____

H. PROGRAM SCHEDULE

This section presents a high-level schedule of the program's deliverables and key milestones.

_____

_____

_____

I. RESOURCE REQUIREMENTS

This section states the resource requirements for the program and notes when they will be required according to the high-level schedule.

_____

_____

_____

J. RISKS AND ISSUES

This section presents the possible risks and issues that could affect the program, positively or negatively, that have been identified at the time the program mandate is prepared.

_____

_____

_____

K. STAKEHOLDER CONSIDERATIONS

This section presents a list of those stakeholders that are expected to influence the program, positively or negatively, or have an interest in its outcome.

_____

_____

_____

L. GOVERNANCE REQUIREMENTS

This section states the governance requirements for the program at the program level and also at the component level.

_____

_____

_____

M. APPROVALS

This section contains the approval of the program mandate by the members of the Portfolio Review Board and other key stakeholders as required.

SIGNATURES AND DATE APPROVAL OBTAINED

Portfolio Review Board chairperson _____

Portfolio Review Board member 1 _____

Portfolio Review Board member 2 _____

Portfolio Review Board member 3 _____

Stakeholder 1 _____

Stakeholder 2 _____

Stakeholder N _____

Program Roadmap

Timing of major milestones is important in the selling and execution of a program. The program roadmap provides a guide or "map" through time of the program's scope and execution. There is considerable similarity of the program roadmap to the concept of the integrated master plan in the United States Department of Defense (2005). At this time the roadmap is high level in nature. It is greatly refined later in the program, especially as the program management plan is prepared.

The terms roadmap and timeline are sometimes confused. Indeed, a roadmap document should include a (usually) single graphic depiction of its essence. That depiction is a timeline, but in the roadmap there is far more strategic information. The program roadmap graphic not only conveys the essence of the program in a single diagram, but it also must convey that information completely enough in a manner readily assimilated by higher-level executives in the organization.

Program Roadmap Instructions

The program roadmap includes the following:

Purpose: A brief introductory statement defining the purpose of the program roadmap, such as:

> The program roadmap shows, in a chronological way, the program's intended direction as it describes major milestones, key dependencies, the link between the planned and prioritized work, and key decision points.

The program roadmap helps in program execution and management, as it is used to assess progress in delivering benefits. It shows the program's high-level, overall scope and execution. It is prepared based on the program's business case and the organization's strategic objectives.

Although the program roadmap is prepared as the program is initiated, it is an iterative document, developed in a rolling-wave format, and it should be reviewed periodically by the program management team, other key stakeholders, and the members of the Governance Board, as the work of the program continues throughout the various phases in its life cycle, especially when there are schedule, financial, benefit, and other program changes.

The program roadmap is a subsidiary document to the program management plan and evolves with it. It is referenced often in the *Standard for Program Management*—Third Edition (2013). Please see section 3.2, which points out its usefulness in showing milestones details, deliverables, assumptions, and benefits, pointing out it is essential in executing the program and positioning it for success. Section 4.2.2 describes its importance in benefit management in showing objectives, challenges, and risks. It also is described, along with the program charter, in Section 8.3.1.6, with a focus on communicating the link between the business strategy and the planned and prioritized work.

Endpoint objectives: This section describes the overall objectives of the program, especially in terms of benefits realization and ultimate benefits sustainment. It describes the link between the business strategy and the planned and prioritized work to be done in the program. It may include any needed resources or competencies required for program success.

Key challenges and risks: This section describes the key challenges associated with the program and the identified risks to the program. By noting these challenges and risks as part of the roadmap, deviations from them then may indicate other threats to the program or potential opportunities.

Key milestones and decision points: This section describes the key milestones and decision points in the program. As part of the roadmap, the high-level, overall scope and execution of the program are shown. These milestones and key decision points will increase as the program ensues, as components are completed, and others are initiated as part of the program. As decisions are made, and as change requests from other areas cause milestones to change, the roadmap may require updates. This section may include a graphic roadmap fashioned after the simple example at the end of this section. The configuration of that depiction is highly dependent on the individual program and must be tailored as such to convey in a single glance the overview of the roadmap document.

Program organization: This section describes the dependencies between the program components, such as releases in a software program or stages in a construction program. It describes when components are to be implemented and how requests to initiate components are processed.

Supporting infrastructure: This section describes the support structure and needed capabilities for the program, such as use of a program management office; guidelines, procedures, and templates; facilities; and tools and techniques.

Approvals: This section contains the written approval of the program roadmap by the program sponsor, program manager, program management office, members of the Governance Board, and any other key stakeholders as appropriate.

Program Roadmap Template

<Insert Program Name>
Program Roadmap

Program name:	
Program manager:	PM's email address here as a hyperlink
Program sponsor:	
Actual start date:	
Approved end date:	
Program number:	
Revision history:	
Business unit:	

A. PURPOSE

A brief introductory statement defining the purpose of the program roadmap, such as:

> The program roadmap shows, in a chronological way, the program's intended direction as it describes major milestones, key dependencies, the link between the planned and prioritized work, and key decision points.

_____

_____

_____

B. ENDPOINT OBJECTIVES

This section describes the program's overall objectives, especially in terms of benefits realization and ultimately benefits sustainment. It describes the link between business strategy and the planned and prioritized program work.

_____

_____

_____

C. KEY CHALLENGES AND RISKS

This section describes the key challenges associated with the program and identified program risks.

_____

_____

_____

D. KEY MILESTONES AND DECISION POINTS

This section describes the program's key milestones and decision points. It also presents the overall scope and execution of the program. This section should be supplemented by a graphical depiction (discussed further below) that in some purposes and occasions may often find application as a separate artifact.

_____

_____

_____

E. PROGRAM ORGANIZATION

This section describes the dependencies between the program components, when they will be implemented, and how requests to initiate components are processed.

_____

_____

_____

F. SUPPORTING INFRASTRUCTURE

This section describes the support structure and needed capabilities for the program, such as use of a program management office; guidelines, procedures, and templates; facilities; and tools and techniques.

_____

_____

_____

G. APPROVALS

This section contains the approval of the program roadmap by the program sponsor, program manager, program management office, members of the Governance Board, and other key stakeholders as required.

SIGNATURES AND DATE APPROVAL OBTAINED

Governance Board chairperson _____

Governance Board member 1 _____

Governance Board member 2 _____

Governance Board member 3 _____

Program sponsor _____

Program manager _____

Program management office director _____

Stakeholder 1 _____

Stakeholder 2   _____

Stakeholder N   _____

Program Roadmap Diagramming

The graphical program roadmap can be (and not unusually is) implemented with a variety of tools such as spreadsheets, diagramming tools, scheduling tools, and presentation tools. Each mode has advantages and disadvantages.

It is always convenient when the graphic can be data driven, owing to the fact that many updates can be sometimes made by changing underlying data without as many tedious drawing requirements. Such is the allure of the spreadsheet approach. Spreadsheets at this writing, however, generally do not provide the "plug-in" power to translate into great diagrams, even though spreadsheets can drive presentation tools and diagramming tools.

In terms of design approach, the example program roadmap of section 3.2 of the *Standard for Program Management—Third* Edition (2013) exhibits several good design characteristics:

1. It provides a long-term view of the program's milestones.
2. It gives more detail about near-term milestones.
3. It gives much more condensed information on longer-term milestones.
3. Multiple swim lanes provide a coordinated overview of program areas.
4. Milestones illustrate benefit achievement among other aspects.

See the graphic, which follows, as an approach to consider.

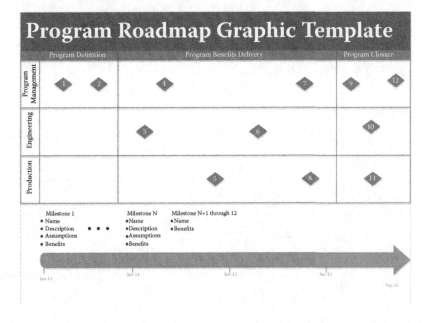

Chapter 4

Program Benefits Management

> Write injuries in dust, benefits in marble.
>
> **—Benjamin Franklin**

The benefits of our problem are the reason we are working on it, and we want them to be not just identified, but realized and sustained so they can be written 'in marble'. Establishing a program indicates that by doing so greater benefits will result than if the projects and other work of the program were managed separately.

On a program, benefits are of many different types. Some may be realized immediately or at various program milestones; others may not be realized until the program is completed or may not even be recognized as benefits until years after the program ends. Some benefits are tangible or easily quantifiable, while others are intangible, difficult to quantify but equally or in some cases even more important. Benefits may be ones of value only to the performing organization or intrinsic, while on other programs they are extrinsic and may be ones used by clients or the society at large.

Within a program, each project or other work has some benefits associated within it that are realized at various times, and they are consolidated as the program ends.

According to the *Examination Content Outline* (PMI, 2011), a high-level view of the key tasks in Benefits Management is as follows:

- Developing a benefits realization plan
- Updating and communicating the plan to stakeholders

29

- Developing a benefits sustainment plan
- Monitoring established metrics to maintain and hopefully improve benefit realization
- Identifying that as projects close the program meets or exceeds its benefit realization criteria
- Maintaining a benefits register
- Analyzing and updating the benefits realization and sustainment plans
- Developing a transition plan to operations to further sustain the program's benefits

The Standard for Program Management—Third Edition (2013) refers to this domain as Program Benefits Management and sets forth a program benefit delivery process, which follows program definition, and ensures that the projects and other work of the program, or its components, are planned, realized, and managed in order that the benefits stated in the initial business case are delivered and possibly exceeded. It entails not only defining the benefits but also creating, realizing, maximizing, and sustaining them. It is therefore a constant theme in program management with a five-stage process of:

- Benefits Identification
- Benefits Analysis and Planning
- Benefits Delivery
- Benefits Transition
- Benefits Sustainment

Key artifacts in Program Benefits Management follow, building on the Initial Program Benefits Statement prepared in Strategic Program Management/Alignment.

Benefits Breakdown Structure

While a Program Work Breakdown Structure is a recognized best practice in program management, a Benefits Breakdown Structure also is useful. Thiry (2004) described a functional diagramming method, which in use of programs he called it a Benefit Breakdown Structure.

Its purpose is to show how the program's vision or desired end state is realized through strategic objectives that are implemented through benefits until each project's or non-project work in the program produce capabilities, which then contribute to business benefits, the strategic objectives, and the overall vision (Thiry, 2010).

Benefit Breakdown Structure Instructions

The benefits breakdown structure builds on the initial benefits statement described in the previous chapter, and the benefits listed in the business case. It is useful in

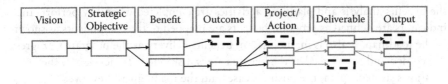

Figure 4.1 Benefits Breakdown Structure example.

benefits Identification to ensure the program manager and team members have not overlooked any benefits expected from the program. It is essential to define the benefits to qualify their business value and then to determine how best to measure their achievement.

To develop the benefits breakdown structure, a similar approach to developing a program work breakdown structure (PWBS) can be used following the decomposition process.

Start with the overall program and its strategic objectives from the business case. Then follow these steps:

1. List the projects and other work to be performed by the program.
2. Identify the specific benefits that each project or operational activity is to realize.
3. Identify program-level benefits that are not associated with any of these components.
4. Each identified benefit then is a benefit package.
5. Link the benefit package to the appropriate PWBS element.
6. Assign the responsibility for benefit package to an individual or organizational unit.

Figure 4.1 shows a graphical representation of the benefit breakdown structure.

Benefits Register

As the benefits are identified, a benefits register becomes a critical program decision-making tool that is updated throughout the program. It is first prepared as part of benefits identification according to *The Standard for Program Management—Third Edition* (2013). As more detail is known about the program's benefits, especially in reviews with stakeholders and members of the program's Governance Board or comparable group, additional details are added to it to describe how each benefit will be measured, evaluated to assess its achievement, the date the benefit is to be realized, and the responsible person, organizational unit, or group. It is used by the Governance Board to assist in evaluating the status of the program's benefits and as a way to communicate their achievement to program stakeholders.

Therefore, as noted in *The Standard for Program Management*—Third Edition (2013), it assists in developing the benefits realization plan and is updated after the benefits analysis and planning activities are completed when key performance indicators are defined as program benefits are mapped to the components as shown in the roadmap. It also is maintained in benefits delivery as the program's progress is recorded and is reviewed as the benefits are transitioned. To develop it, the organization's strategic plan, the business case, and the roadmap are reviewed.

Benefits Register Instructions

The benefits register is prepared as a table and a description of its contents follows:

1. **Benefits identification number:** Assign an identification number to the benefit.
2. **PWBS number:** Link the benefit to the corresponding program package in the program work breakdown structure.
3. **Benefits breakdown structure number:** Link the benefit to the corresponding benefit package in the benefit breakdown structure.
4. **Benefit description:** Describe the benefit and its significance to the program.
5. **Benefit type:** Describe the type of benefit such as:
   - Financial or non-financial
   - Tangible or intangible
   - Extrinsic or intrinsic
6. **Benefit owner:** List the person, organizational unit, or department responsible for realization of the benefit with contact information.
7. **Program contribution:** Determine the estimated contribution of the benefit to the program in terms of the strategic objectives.
8. **Benefit realization:** State how the benefit is to be realized.
9. **Resources required:** List the resources required to realize the benefit.
10. **Planned realization date:** State the date the benefit is to be realized.
11. **Actual realization date:** State the date the benefit was realized; provide information as needed if the dates are different and describe any corrective actions taken.
12. **Measurement criteria:** Consider the type of benefit and describe how the benefit will be measured such as:
    - Financial analysis
    - Earned benefit
    - Hours saved
    - Increased market share
    - Improved productivity
    - Increase in customer satisfaction
    - Decrease in personnel turnover
    - Reduction in the strength of competitors
13. **Measurement method:** List how often measurements will be taken and the estimated effort involved.
14. **Risks:** List any possible risks that may affect benefit realization.

15. **Stakeholders:** List the key stakeholders or stakeholder groups interested in the benefit and their contact information.
16. **Approved by:** State who approved the benefit register. Examples include:
    - Program sponsor
    - Governance Board members
17. **Notes:** Use this field for any additional information about the benefit.

Benefits Realization Plan

The benefits realization plan is the "one-stop shop" for benefits identification, monitoring, and implementation of benefits achievement, and defining them in terms of measurable program outcomes.

In the Project Management Institute's *Standard for Program Management—Third Edition* (2013), the benefits realization plan is noted in the benefits management life cycle under phase two, benefits analysis and planning. During Strategic Program Management, the importance of benefits is noted as part of the business case used to justify the program (3.1.1) and to determine whether or not it should be part of the organization's portfolio. Benefits also are described in the business case for each of the projects in the program. The program roadmap (3.1.2) shows the linkage between program activities and expected benefits.

The benefits realization plan is noted in Section 4.2 and serves to guide the work done in the program. It is a significant document as it serves as a baseline. After it is prepared, the benefit register should be updated (4.2.3). The plan is used in benefits delivery (4.3) and its sub-sections, program benefits and program components (4.3.1), program benefits and program governance (4.3.2), and may need modifications. It is also used in program management plan development (8.3.2) and in scope definition in program scope management (8.9). If the benefits realization plan is modified, the roadmap requires updating (4.3.2).

Benefits Realization Plan Instructions

The benefits realization plan includes the following:

Purpose: A brief introductory statement defining the purpose of the benefits realization plan, such as:

> The benefits realization plan assesses the value and organizational impact of the program. It identifies the benefits to be realized from the projects and other work of the program and ensures the benefits are specific, measurable, attainable, realistic, and time based.
>
> It also analyzes the impact of changes on the overall outcome of the program and assigns roles and responsibilities to ensure the identified benefits are

attained. Although the plan is drafted early in the program, it is an iterative document that should be further refined as the program progresses through the life cycle. This plan is included in the program management plan as it helps to determine how benefits will be realized and provides a baseline for tracking progress and reporting variances. It also focuses on an effective transition to an operation state once the program ends.

Definition of each program benefit: This section defines each benefit in the program—tangible and intangible, financial and non-financial, or extrinsic or intrinsic. It also shows how the benefit is to be realized. The benefits described in this section should be in line with those in the program's business case and the organization's overall strategy. Benefits are identified through interviews, brainstorming sessions, and review sessions. It builds on the benefits already identified in the benefit register.

Assumptions: This section lists any assumptions for the benefits that have been identified and defines them further in this plan. Assumptions are facts that the team considers during planning to be true, real, or certain. In terms of benefits, examples include: availability of required resources in terms of the schedule to realize each benefit, measuring the status of each benefit defined in the plan, and identifying opportunities to realize even more benefits than those initially defined.

Each benefit's impact to the program's outcomes: This section shows how each benefit relates to the outcomes of the program. It further shows the value of the program in terms of the benefits to be realized, and it describes the interdependencies between the program's benefits.

Metrics and procedures to measure benefits: In order to ensure that the program is realizing its intended benefits, this section describes the key metrics that will be collected, and the processes that will be used to measure the program benefits. Suggested metrics include scheduled time for benefits realization versus actual time, extent of the benefits realized versus those planned in the business case, impact of the realized benefit on other components of the program, the need to change the benefits as the program progresses versus those in the plan, the value of the benefits that were realized, the extent of the sustainment of benefits after the program ends, and other benefits realized that were not planned. As well, this section includes a process to determine the extent to which each benefit has been achieved before the program is closed. The metrics can be used if the benefits are not realized as planned in order that corrective actions can be implemented.

Roles and responsibilities for benefits realization and management: Benefits realization and management is a team effort. This section describes specific roles and responsibilities of the program team members to deliver final and intermediate benefits in the program. These roles and responsibilities can be displayed in a Responsibility Assignment Matrix (RAM) or in a Responsible, Accountable, Consult, Inform (RACI) chart to show whether the team member is responsible for the benefit, should be involved in its realization in terms of monitoring and control, should be performing an ongoing analysis of the program for incremental benefits to make adjustments to the benefits delivery schedule, should be consulted as the benefit is realized, or should sign off once the benefit has been realized.

Benefits management communications plan: Program stakeholders require information concerning how the proposed program benefits are being realized. This section describes the communications methods to be used to ensure each stakeholder has the information he or she requires regarding benefits management. It also describes how benefits will be reviewed with stakeholders. It discusses the preparation on a regular basis of the benefits realization report, since this report is an output of the report performance process, and describes the benefits realization plan versus the actual benefits delivered and serves to monitor overall benefits delivery. The benefits management communications plan can be an appendix or section in the program's communications management plan.

Benefits delivery schedule: For many programs benefits will not be realized at the same time; however, in others they will be realized incrementally. Also, some projects in the program will need to integrate the benefits from their projects with other projects or ongoing work of the program. This section presents a schedule, typically in a Gantt or network chart format, to show the planned and actual dates for each of the identified benefits in the program. The schedule should show the interdependencies between benefits from the various projects in the program and the other work that is part of the program. This schedule should be part of the program's master schedule. It also is included as part of the program roadmap.

Benefits-related risks: Each benefit will have some type of risks associated with its realization. This section describes the risks and stresses the risks that may occur based on the component risks on benefits delivery so they can be evaluated.

Required changes to processes and systems: This section discusses new processes and systems and those already in existence, which must be changed in order to track and monitor benefits as they are being realized in the program.

Transition of the program's benefits into ongoing operations and benefits sustainment: When the program is officially closed, the program is transitioned into ongoing operations. This section describes how the program will transition and how the program's benefits will be sustained. It is important to ensure that the transition activities provide for continued management of the benefits within the framework of the customer organization as appropriate. Recognize that benefits management transcends the program's life cycle and continues into transfer and sustainment of the benefits of the program.

Approvals: This section contains the written approval of the benefits realization plan by the program sponsor, program manager, program management office director, members of the Governance Board, and other stakeholders.

Benefits Realization Plan Template

<Insert Program Name>
Benefits Realization Plan

Program name:	
Program manager:	PM's email address here as a hyperlink
Program sponsor:	
Actual start date:	
Approved end date:	
Program number:	
Revision history:	
Business unit:	

A. PURPOSE

A brief introductory statement defining the purpose of the benefits realization plan, such as:

> The benefits realization plan assesses the value and organizational impact of the program. It identifies the benefits to be realized from the projects and other work of the program, and ensures the benefits are specific, measurable, attainable, realistic, and time based.

_____

_____

_____

B. DEFINITION OF EACH PROGRAM BENEFIT

This section defines each benefit in the program—tangible and intangible, financial and non-financial, extrinsic or intrinsic. It also shows how the benefit is to be realized.

_____

_____

_____

C. ASSUMPTIONS

This section lists any assumptions associated with the benefits defined in the plan such as availability of resources at the required time.

_____

_____

_____

D. EACH BENEFIT'S IMPACT TO THE PROGRAM'S OUTCOMES

This section shows how each benefit relates to the outcomes of the program. It further shows the value of the program in terms of the benefits to be realized.

_____

_____

_____

E. METRICS AND PROCEDURES TO MEASURE BENEFITS

This section describes the key metrics that will be collected, and the processes that will be used to measure the program benefits.

_____

_____

_____

F. ROLES AND RESPONSIBILITIES FOR BENEFITS REALIZATION AND MANAGEMENT

This section describes specific roles and responsibilities of the program team members to deliver the final and intermediate benefits in the program. Attach a Responsibility Assignment Matrix (RAM) or a Responsible, Accountable, Consult, Inform (RACI) chart as appropriate.

_____

_____

_____

G. BENEFITS MANAGEMENT COMMUNICATIONS MANAGEMENT PLAN

This section describes the communications methods to be used to ensure each stakeholder has the information he or she requires regarding benefits management. It also describes how benefits will be reviewed with stakeholders and the content of the benefits realization report.

_____

_____

_____

H. BENEFITS DELIVERY SCHEDULE

This section presents a schedule, typically in a Gantt or network chart format, to show the planned and actual dates for each of the identified benefits. The schedule should show the interdependencies between benefits from the various projects in the program and other work that is part of the program. The benefits as well should be part of the program roadmap.

_____

_____

_____

I. BENEFITS-RELATED RISKS

This section describes the risks associated with benefits realization.

_____

_____

_____

J. REQUIRED CHANGES TO PROCESSES AND SYSTEMS

This section discusses new processes and systems and those already in existence that must be changed to handle benefits realization and management.

_____

_____

_____

K. TRANSITION OF THE PROGRAM'S BENEFITS INTO ONGOING OPERATIONS AND BENEFITS SUSTAINMENT

This section describes how the program will transition into ongoing operations with an emphasis on how the program's benefits will be sustained.

_____

_____

_____

L. APPROVALS

This section contains the approval of the benefits realization plan by the program sponsor, program manager, program management office, members of the Governance Board, and other key stakeholders.

Program sponsor _____

Program manager _____

Program management office director _____

Governance Board chairperson _____

Governance Board member 1 _____

Governance Board member N _____

Stakeholder 1 _____

Stakeholder 2 _____

Stakeholder N _____

Benefits Realization Report

This report specifies progress toward achieving benefits and correlates related expenditures. While this report is not mentioned in *Standard for Program Management*—Third Edition (2013), we have retained it from the *Standard for Program Management*—Second Edition (2008) and have modified it slightly because of the importance of benefits to program management.

Benefits Realization Report Instructions

The benefits realization report includes the following:

Purpose: A brief introductory statement defining the purpose of the benefits realization report, such as:

> The benefits realization report describes the benefits realized to date to ensure that they are aligned with the program's business case and benefits realization plan.

> Programs are established to achieve benefits that may not be realized if its components were managed individually. Benefits realization and effective benefits transition and sustainment are keys to a successful program. Benefits have value when they are used by the customer, by an operations or support group, or by another program in the organization. Programs are interested in the progress of the program in terms of its realization of benefits.

Since the program components will deliver benefits at different times, the benefits realization report needs to be prepared and issued on a regular basis so stakeholders are informed of progress and to see whether any corrective or preventive actions are required. The benefits realization report should track to the benefits realization plan.

The following are the recommended contents for the benefits realization report:

Background information:
> **Program overview:** This section presents a brief overview of the program in terms of the benefits that are expected to be realized based on the benefits realization plan and to ensure that the benefits remain aligned with the business case.
>
> **Reporting period:** This section states the reporting period for the report. A monthly report is recommended, with ad hoc communications as required when benefits are achieved before the report is formally issued.
>
> **Prepared by:** This section provides contact information for the program management team member who prepared the report.

Benefits realized during the reporting period:
> **Quantitative benefits:** This section states the quantitative benefits that were realized during the reporting period in conjunction with the program's business case and its objectives. It describes when the benefits realization starts in terms of transfer to ongoing operations, the customer, or another program. It also states quantifiable measures of success, such as an increase in productivity, market share, or profits; improvements over those of the competition; net present value; shareholder value added, return on investment; benefit-cost; extent of rework required; product conformance with requirements; and customer acceptance of deliverables.
>
> **Qualitative benefits:** This section states qualitative benefits such as customer satisfaction, team satisfaction, use of processes and procedures, value of program management tools used, strategic importance, impact of the deliverables on the organization, improvements in morale, validity of the program vision, effect of technology in terms of benefit achievement, the involvement of team members in performance improvement initiatives, or a societal benefit such as reduced exposure to a foodborne health hazard. It is necessary to then describe these qualitative benefits in quantitative terms.

Benefits not realized during the reporting period: This section states those benefits that were expected to be realized during the reporting period but were not completed according to the benefits delivery schedule in the benefits realization plan. It describes any corrective actions that are required in order to meet the delivery schedule so that actions can be taken for successful benefits delivery. Corrective action may lead to the need to add new components, make revisions to existing components, or terminate a component.

New benefits: This section states any new benefits from existing components that now have led to a need to revise the benefits realization plan and the program roadmap. It describes any interdependencies between these new benefits and those already identified in the plan. This section also states who is responsible to ensure that these benefits are realized and delivered as planned.

Approvals: This section contains the written approval of the benefits realization report by the program manager, program sponsor, program management office director, members of the Governance Board, and any other key stakeholders as appropriate.

Benefits Realization Report Template

Background Information		
Program Overview	*Reporting Period*	*Prepared by*

Benefits Realized during the Reporting Period	
Quantitative Benefits	
Benefit Description	*Date Realized*

Qualitative Benefits	
Benefit Description	*Date Realized*

Benefits Not Realized during the Reporting Period	
Benefit Description	*Planned Corrective Action*

continued

Benefits Realization Report Template (continued)

New Benefits	
Benefit Description	*Assigned to*
	.
Approvals	
Program manager	Date
Program sponsor	Date
Program management office director	Date
Governance Board chairperson	Date
Governance Board member 1	Date
Governance Board member 2	Date
Governance Board member N	Date
Stakeholder 1	Date
Stakeholder 2	Date
Stakeholder N	Date

Program Benefits Transition Plan

Without transition, program benefits are, in most cases, simply forgotten or terminated. Without proper transition management for both the components and the program, the program lacks value. As noted in the *Standard for Program Management*—Third Edition (2013), value from the program is only possible if the beneficiaries can use its benefits.

The scope of the benefits transition must be defined, and a best practice to follow is to involve the key stakeholders who will be responsible for the transition in preparing this plan. The delivered program benefits are measured and consolidated, and the benefits register (4.1.2) is updated. The realized benefits should be compared to those in the business case (4.1.1). It is recognized that other activities are part of the program transition process. The importance of benefit transition and sustainment also is discussed as part of the closing processes (8.3.6).

Program Benefits Transition Plan Instructions

The program benefits transition plan will include the following:

Purpose: A brief introductory statement defining the purpose of the program benefits transition plan, such as:

> The program benefits transition plan describes the process that will be followed to facilitate the ongoing realization of the program's benefits when the benefits are delivered to the customer; to another unit in the performing organization, such as a product support, customer support, or service management group; or to another program that is under way or about to start to ensure the benefits can be sustained.

> This plan serves to provide a smooth transition process. The objective is to ensure that the benefits from the program continue to be realized as long as possible. Once the transition activities are complete, program resources then are reallocated, and all program records are closed. It is an iterative document that may be refined as various program components are closed and transitioned to ongoing operational status or to the customer.

> In the Project Management Institute's *Standard for Program Management—Third Edition* (2013), the benefits transition plan is the fourth phase in the benefits life cycle. It also notes that programs may be terminated without a transition to operations especially if the program charter has been fulfilled or if the program no longer is valuable to the organization.

Transition process: This section describes the benefits transition process to be followed. It states the key stakeholders involved in preparing the plan, ways to ensure resulting benefits are measured, the benefits register is updated, acceptance criteria from the results or outputs from program components are met, and approval requirements. The receivers need to be ready to accept the benefits, which may require training, workshops, meetings, job aids, support systems, complete documentation, etc. A schedule for the benefit transition activities should be part of the overall program master schedule and included in the program roadmap.

Component transition requests: Throughout the program, different components (projects and non-project work) will finish before others. Each time a component is officially closed, a transition request should be prepared to transition the benefits from the component into ongoing operational status, to the customer, or to another program. This section describes the process to prepare these requests, the format to be used, and the approval requirements. Typically, these activities are part of the last gate review for the component.

Documentation requirements: This section describes documentation to be prepared as part of the benefit transition process. It also includes a description of any training to be provided or materials that may be needed. Often, transitions are formal contract-type activities; in other situations, a more informal process may be appropriate. Meetings and conferences may be required. While the receivers will vary based on the component and type of program, a clear understanding of the specific benefits to be transitioned is required so there are no misunderstandings later in the process.

Roles and responsibilities: This section describes the roles and responsibilities of the program management team as well as the receiving organization in the transition process. The receiving organization ideally should be a participant in preparing the transition plan so there is a clear understanding of what will be handed to this unit, and the process that will be followed.

Transition critical success factors: This section describes the critical success factors that are considered essential to ongoing benefits transition and sustainment. It may include items such as continuing to meet product demands, adding value to the ongoing process that will be followed, having a customer or product support representative available, providing an assessment later to determine if the resulting changes have been implemented successfully, providing support if changes are required, disposing and reallocating required resources, and training support staff as needed.

Approvals: This section contains the written approval of the program benefits transition plan by the program sponsor, program manager, program management office, members of the Governance Board, and other stakeholders.

Program Benefit Transition Plan Template

<Insert Program Name>
Program Benefit Transition Plan

Program name:	
Program manager:	PM's email address here as a hyperlink
Program sponsor:	
Actual start date:	
Approved end date:	
Program number:	
Revision history:	
Business unit:	

A. PURPOSE

A brief introductory statement defining the purpose of the program benefits transition plan, such as:

> The program benefits transition plan describes the process that will be followed to facilitate the ongoing realization of the program's benefits to the customer or to transition it to another unit in the performing organization, such as a product support, customer support, or service management group; or to another program in the organization; or to a program about to be started to ensure program benefits are sustained.

B. TRANSITION PROCESS

This section describes the program benefit transition process to be followed. It states the key stakeholders involved in preparing the plan, ways to ensure resulting benefits are measured, the benefits register is updated, and approval requirements.

C. COMPONENT TRANSITION REQUESTS

This section describes the process to be followed to prepare a transition request each time a component is officially closed to transition the benefits from the component into ongoing operational status, to the customer, or to another program.

D. DOCUMENTATION REQUIREMENTS

This section describes the documentation to be prepared as part of the transition process. It also includes a description of any training to be provided and materials that may be needed.

E. ROLES AND RESPONSIBILITIES

This section describes the roles and responsibilities of the program management team as well as the receiving organization in the transition process.

_____

_____

_____

F. TRANSITION CRITICAL SUCCESS FACTORS

This section describes the critical success factors that are considered essential to ongoing benefits transition and sustainment.

_____

_____

_____

G. APPROVALS

This section contains the approval of the program transition plan by the program sponsor, program manager, program management office, members of the Governance Board, and other key stakeholders.

SIGNATURES AND DATE APPROVAL OBTAINED

Program manager   _____

Program sponsor   _____

Program management office director   _____

Governance Board chairperson   _____

Governance Board member 1   _____

Governance Board member 2   _____

Governance Board member N   _____

Stakeholder 1   _____

Stakeholder 2   _____

Stakeholder N   _____

Program Benefits Sustainment Plan

Having gone through transition, benefits should be ready for continuing accrual. While the benefits transition plan is intended to provide necessary forethought and preparation for ongoing benefits accrual, in and of itself it does not provide all of the detailed mechanisms, processes, and changes necessary for indefinite success as the transitioned benefits continue in a changing world. According to the *Standard for Program Management*—Third Edition (2013, the plan should address the risks, processes, measures, metrics, and tools that will be needed.

Benefits Sustainment Plan Instructions

The benefits sustainment plan includes the following and must be a living document continually adapting to its environment:

Purpose: A brief introductory statement defining the purpose of the benefits sustainment plan, such as:

> The benefits sustainment plan provides the mechanisms necessary to assure that benefits continue to accrue in an ever changing environment. It draws extensively on the contents of the benefits transition plan.

Definition of each program benefit: Taken from program documentation and specifically the benefits register, this section defines for reference the program benefits to be sustained.

Sustainment critical success factors: This section describes the critical success factors that are considered essential to ongoing benefits sustainment. It may include items such as continuing to meet product demands, adding value to the ongoing process that will be followed, having a customer or product support representative available, providing an assessment later to determine if the resulting changes have been implemented successfully, providing support if changes are required, disposing and reallocating required resources, and training support staff as needed.

Assumptions: This section lists any assumptions for the benefits that have been identified previously or are further identified defined in this plan. Assumptions are facts that the team considers during planning to be true, real, or certain.

Metrics and procedures to measure benefits: In order to ensure that the program is realizing its intended benefits, this section describes the key metrics that will be collected, and the processes that will be used, to measure the program benefits. Suggested metrics include scheduled time for benefits sustainment versus actual time, extent of the benefits realized versus those planned in the business case, impact of the realized benefits on other components of the program, the need to change the benefits as the transitioned program progresses versus those in the plan, the value of the benefits that were realized, the extent of the sustainment of benefits after the program ends, and other benefits realized that were not planned. These metrics can be used if the benefits are not realized as planned in order that corrective actions can be implemented.

Roles and responsibilities for benefits sustainment: Benefits sustainment and management is a team effort and must address all critical success factors. This section describes specific roles and responsibilities of the transitioned team members to deliver final and intermediate benefits in the program. These roles and responsibilities can be displayed in a responsibility assignment matrix (RAM) or in a Responsible, Accountable, Consult, Inform (RACI) chart to show whether the team member who is responsible for the benefit should be involved in its sustainment in terms of monitoring and control, should be performing an ongoing analysis of the program for incremental benefits to make adjustments to the benefits delivery schedule, should be consulted as the benefit is realized, or should sign off once the benefit has been sustained.

Benefits sustainment communications plan: Program stakeholders require information concerning how the proposed program benefits are being sustained. This section describes the communications methods to be used to ensure each stakeholder has the information he or she requires regarding benefits sustainment and identifies the interested stakeholders.

Benefits sustainment-related risks: Each benefit will have some type of risks associated with its ongoing successful sustainment. This section describes the risks and stresses the risks that may occur.

Required changes to processes and systems: This section discusses new processes and systems and those already in existence, which must be changed in order to track and monitor benefits as they are being realized in the program.

Approvals: This section contains the written approval of the benefits sustainment plan by the program sponsor, program manager, program management office director, members of the Governance Board, and other stakeholders.

Benefits Sustainment Plan Template

<Insert Program Name>
Benefits Sustainment Plan

Program name:	
Program manager:	PM's email address here as a hyperlink
Program sponsor:	
Actual start date:	
Approved end date:	
Program number:	
Revision history:	
Business unit:	

A. PURPOSE

A brief introductory statement defining the purpose of the benefits sustainment plan, such as:

> The benefits sustainment plan provides the mechanisms necessary to assure that benefits continue to accrue in an ever changing environment. It draws extensively on the contents of the benefits transition plan.

_____

_____

_____

B. DEFINITION OF EACH PROGRAM BENEFIT

For reference in sustainment, this section defines each benefit in the program—tangible and intangible, financial and non-financial, extrinsic or intrinsic. It also shows how the benefit is to be realized.

_____

_____

_____

C. SUSTAINMENT CRITICAL SUCCESS FACTORS

This section describes the critical success factors that are considered essential to ongoing benefits sustainment.

_____

_____

_____

D. ASSUMPTIONS

This section lists any assumptions associated with the benefits defined in the plan such as availability of resources at the required time.

_____

_____

_____

E. METRICS AND PROCEDURES TO MEASURE BENEFITS

This section describes the key metrics that will be collected, and the processes that will be used to measure the program benefits.

F. ROLES AND RESPONSIBILITIES FOR BENEFITS SUSTAINMENT

This section describes specific roles and responsibilities of the transitioned personnel to sustain the ongoing benefits. Attach a Responsibility Assignment Matrix (RAM) or a Responsible, Accountable, Consult, Inform (RACI) chart as appropriate.

G. BENEFITS SUSTAINMENT COMMUNICATIONS PLAN

This section describes the communications methods to be used to ensure each stakeholder has the information he or she requires regarding benefits sustainment. It also describes how benefits will be reviewed with stakeholders and describes the content of the resulting report(s).

H. BENEFITS SUSTAINMENT-RELATED RISKS

This section describes the risks associated with benefits sustainment.

I. REQUIRED CHANGES TO PROCESSES AND SYSTEMS

This section discusses new processes and systems and those already in existence that must be maintained to handle benefits sustainment and management.

J. APPROVALS

This section contains the approval of the benefits sustainment plan by the program sponsor, program manager, program management office, members of the Governance Board, and other key stakeholders.

SIGNATURES AND DATE APPROVAL OBTAINED

Program sponsor   _____

Program manager   _____

Program management office director   _____

Governance Board chairperson   _____

Governance Board member 1   _____

Governance Board member N   _____

Stakeholder 1   _____

Stakeholder 2   _____

Stakeholder N   _____

Chapter 5

Program Stakeholder Engagement

> The greatest ability in business is to get along with others and to influence their actions.
>
> —John Hancock

Program managers must be outstanding communicators as they deal with numerous stakeholders at all levels of the organization and with external stakeholders as well. Research has shown for years that project managers spend about 90% of their time communicating. Given the complexity of programs and the larger number of stakeholders, it is evident that the program manager spends even more time communicating with stakeholders on most programs.

The proactive program manager is one who strives to identify stakeholders who have an interest in or can influence his or her program as early as possible and then to set forth to engage them effectively. The objective is for all stakeholders to be positive and supportive of the program's goals and to work collaboratively to achieve them. While it may not be possible for everyone to be totally supportive, through effective identification and ongoing communications, hopefully the program manager and his or her team can work with any stakeholders who are negative, determine why they are not supportive, and address their concerns as much as possible so that, at a minimum, they become neutral toward the program. One negative and outspoken stakeholder can have the potential to ensure the program does not meet its goals and achieve its benefits. Ongoing and proactive stakeholder engagement is critical.

According to the *Examination Content Outline* (PMI, 2011) a high-level view of the key tasks in Stakeholder Management is as follows:

- Identifying stakeholders
- Performing stakeholder analysis
- Negotiating for stakeholder support
- Generating and maintaining program visibility for stakeholder support to achieve the program's strategic objectives
- Defining and maintaining communications adapted to the needs of different stakeholders to ensure their support
- Evaluating any risks stakeholders identify and incorporating them into the program's risk management plan
- Developing and fostering stakeholder relationships thereby improving communications and enhancing their support

The Standard for Program Management—Third Edition (2013) refers to this domain as Program Stakeholder Engagement since management of stakeholders is difficult at best. Stakeholder engagement emphasizes capturing stakeholder's needs, desires, and expectations, analyzing the program's impact on its stakeholders, obtaining and maintaining stakeholder support, managing communications, and mitigating resistance. Since programs result in change, stakeholder commitment and involvement in the program are essential, especially since some stakeholders are the beneficiaries of the program's benefits. It is a continual theme in program management consisting of:

- Program Stakeholder Identification
- Program Stakeholder Engagement Planning
- Program Stakeholder Engagement

Key artifacts in Program Stakeholder Engagement follow.

Stakeholder Register

As stakeholders are identified, listing them in a stakeholder register and maintaining this register is a best practice to follow. It can be done in the stakeholder register whether in a spreadsheet or database form. As it may contain some sensitive information, the program manager should determine the people who have access to it and may decide to omit certain information from widespread distribution. This register is prepared in program stakeholder identification in the *Standard for Program Management*—Third Edition (2013).

Stakeholder Register Instructions

Purpose: The stakeholder register is one of the first documents prepared in the program. An initial list of stakeholders is in the business case for the program, is refined during the planning process as the stakeholders are identified, and then is updated as the program planning continues and during program execution. Parts of it can be an appendix to the stakeholder engagement plan and also to the program communications management plan. It is designed as a table, and a description of its contents follows.

1. **Name:** List the name of the stakeholder.
2. **Position:** It is important to note that some stakeholders may not be aware of the program nor may be supportive of it. List the stakeholder's position in the program. Typical examples are:
   a. Program director
   b. Program manager
   c. Project managers
   d. Program sponsor
   e. Customers
   f. Potential customers
   g. Program team members
   h. Project team members
   i. Functional/department managers
   j. Funding organization
   k. Program management office director
   l. Portfolio manager
   m. Members of the Program Governance Board
   n. Suppliers
   o. Government regulatory agencies
   p. Competitors
   q. Groups (environmental, consumer, or other types of groups affected by the program)
   r. Others (people in other programs or other organizational units that may require similar resources or may have goals that conflict with or complement those of the program; they may benefit from the program or be at a disadvantage because of it)
3. **Contact information:** List contact information for each stakeholder: phone numbers (office and cell), location, and email. For government agencies, competitors, and groups, list the name and contact details of the principal representatives with an interest in the program.
4. **Areas of influence:** State whether the stakeholder has a high level of influence concerning the program. It typically is based on the stakeholder's level of power and authority. Consider a ranking scale such as the following:

a. 5 = Major influence throughout the program life cycle; should be consulted regularly and should meet regularly with the program management team.

b. 4 = Involved at certain phases during the program; should be informed of progress regularly.

c. 3 = Involved at only one phase of the program's life cycle; should be informed of progress regularly in the area of interest but infrequently at other times.

d. 2 = Somewhat interested in the program; should be kept informed of the progress of the program from time to time.

e. 1 = Requires limited notification as to the progress of the program.

5. **Program opinion:** List whether the stakeholder is considered to be a strong supporter of the program, a neutral party, or someone who does not believe the program should be pursued. Consider a ranking scale such as the following:

   5 = Strong program advocate
   4 = Supporter
   3 = Neutral
   2 = Limited interest in the program
   1 = Does not believe the program should be pursued

6. **Program impact:** State whether the stakeholder will be directly or indirectly impacted by the outcome of the program. Consider a ranking scale such as the following:

   5 = Direct impact
   4 = Somewhat impacted
   3 = Neutral
   2 = Limited impact
   1 = No impact

7. **Management strategy:** State the approach the program management team plans to use to work with this stakeholder. Consider a RACI chart such as the following:

   A = Approves all key documents and decisions as required
   C = Consulted before a decision is made and receives a preliminary draft of documents before they are in final form
   I = Informed of progress and can receive documents upon request
   R = Responsible for a critical area in the program

8. **Information requirements:** State the stakeholder's information requirements. Consider an approach such as the following:

   5 = Receives all issued reports and requests others as needed
   4 = Receives all issued reports
   3 = Receives monthly reports
   2 = Receives quarterly reports
   1 = Receives press releases or general program information

9. **Responsible team member:** State the name and contact information of the program team member who will be the principal point of contact for work with this stakeholder.
10. **Issues identified:** List any issues this stakeholder identifies during the course of the program and the date.
11. **Resolution and date:** List how these issues were resolved and the date.
12. **Notes:** Use this field for any additional notes about the stakeholder.

In developing the stakeholder register, consider using some of the following approaches:

- Consult your organization's knowledge repository or lessons learned data base for historical Information about stakeholders on other programs in the organization
- Brainstorm potential stakeholders with your team to not only identify stakeholders but also to determine their roles and areas of interest
- Interview your sponsor for suggestions of other stakeholders in addition to those in the business case
- Conduct some interviews using open-ended questions with the key stakeholders identified to get their ideas as to other stakeholders that may have an interest in the program
- Hold a focus group of different types of stakeholders to help identify others
- Consult with other program managers in your organization for suggestions
- Since customers are key stakeholders and many stakeholders can legitimately be viewed as "customers," use customer relationship management (CRM) as an approach to further identify stakeholders and to show their relationship to the program. CRM is widely used to manage interactions with customers, including capture of comprehensive information about them and interaction with and support of them. It could be viewed as "contact management on steroids."
- Send out questionnaires and surveys to additional people and to stakeholders that have been identified to make sure no one is overlooked; through this approach, more people will feel they are involved in the program and will be more committed to it.

A stakeholder analysis also is recommended. Steps to follow in its development are included in the *Standard for Program Management*—Second Edition (2008, p. 230); it is also mentioned in the Third Edition (see 8.1 and 8.2):

1. Obtain an understanding of the organization's culture by using interviews, focus groups, and questionnaires and surveys to help determine stakeholder attitudes and program communication requirements.

2. Determine the stakeholder's degree of opposition or support toward the program's objectives.
3. Evaluate the extent the stakeholder can influence the program's outcomes by evaluating the stakeholder's interest level and ability to impact the program's outcomes.
4. Prioritize stakeholders based on the extent to which they can influence the program's outcomes, positively or negatively.
5. Develop a stakeholder communication's strategy and define methods to best communicate with stakeholders and how often the stakeholders require program information.
6. Develop the stakeholder register.
7. Update the stakeholder engagement plan (to be discussed).
8. Evaluate how receptive the stakeholder is to receiving communications about the program.

Stakeholder Inventory

Since there can be such a large number of stakeholders on programs, the stakeholder inventory can be a useful approach to group these stakeholders into categories and to assist in prioritizing them in terms of their overall impact on the program. While it is not mentioned in the *Standard for Program Management*—Third Edition (2013), it was included in the Second Edition, and we believe it is a useful document for program managers. Similar to the stakeholder register, it also includes sensitive information, and the program manager needs to determine people who can access it. We have modified it slightly for compliance with the Third Edition.

Stakeholder Inventory Instructions

The stakeholder inventory will include the following:

Purpose: A brief introductory statement defining the purpose of the stakeholder inventory, such as:

> The stakeholder inventory is part of the program's commitment to effective stakeholder engagement. It provides a summary of each stakeholder's involvement in the program, his or her possible responses, any identified issues, and possible mitigation or enhancement strategies.

> Different stakeholders will have an interest in or influence over the program at different times. As the business case is prepared, stakeholder identification begins.

Then, after the stakeholder register has been prepared, the stakeholder inventory can follow. It should complement the program stakeholder engagement plan and support the program communications plan. It is an iterative document that should be reviewed periodically during meetings of the program management team and the Governance Board and updated as new stakeholders are identified throughout the program.

Stakeholder categories: To set up the inventory, use categories for the program's stakeholders based on the stakeholders in the register. This section lists the categories that will be part of the inventory, and it also should consider the various roles of the stakeholders, both internal and external.

Stakeholder influence: For each category, determine the stakeholder's level of influence on the program. This section should describe how the stakeholder will be impacted by the program and how the stakeholder will influence it. A matrix can be prepared. A possible example follows the template.

Possible stakeholder responses: Using the categories and the level of stakeholder influence in the program, this section lists possible responses to various changes that may affect the program or specific areas of interest to the stakeholders. For example, some stakeholders may be more concerned about different areas than others, such as:

- Program benefits
- Program deliverables
- Program funding/financial management
- Schedule
- Governance issues
- Communications issues
- Intellectual property
- Vendor issues
- Level of risk
- Issue resolution
- Internal and external influences
- Resources
- Contracts

This section then describes the areas of greatest interest to specific stakeholders by stakeholder category so possible responses can be prepared should a change occur. Negative impacts should be determined in order that the response then can address them quickly to minimize any adverse impacts to the program.

Potential mitigation strategies: This section describes the potential mitigation strategies to consider should an area of interest be affected.

Approvals: This section contains the written approval of the stakeholder inventory by the program sponsor, program manager, program management office director, members of the Governance Board, and others as appropriate.

Stakeholder Inventory Template

<Insert Program Name>
Stakeholder Inventory

Program name:	
Program manager:	PM's email address here as a hyperlink
Program sponsor:	
Proposed start date:	
Proposed end date:	
Prepared by:	
Program number:	
Revision history:	
Business unit:	

A. PURPOSE

A brief introductory statement defining the purpose of the stakeholder inventory, such as:

> The stakeholder inventory is part of the program's commitment to effective stakeholder engagement. It provides a summary of each stakeholder's involvement in the program, his or her possible responses, any identified issues, and possible mitigation strategies.

_____

_____

_____

B. STAKEHOLDER CATEGORIES

This section lists the stakeholder categories that will be part of the inventory.

_____

_____

_____

C. STAKEHOLDER INFLUENCE

This section defines the influence level of each of the stakeholders by category and shows how the stakeholders will be affected by the program.

_____

_____

_____

D. POSSIBLE STAKEHOLDER RESPONSES

This section uses the categories and the stakeholder's influence on the program and lists possible responses to various changes that may affect the stakeholder's specific areas of interest.

_____

_____

_____

E. POTENTIAL MITIGATION STRATEGIES

This section describes the potential mitigation strategies to consider should an area of interest be affected. These strategies are further defined in the stakeholder engagement plan.

_____

_____

_____

F. APPROVALS

This section contains the approval of the stakeholder inventory by the program manager, program sponsor, program management office director, members of the Governance Board, and others as required.

SIGNATURES AND DATE APPROVAL OBTAINED

Program manager _____

Program sponsor _____

Program management office director _____

Governance Board chairperson _____

Governance Board member 1 _____

Governance Board member 2 _____

Governance Board member N _____

Stakeholder 1 _____

Stakeholder 2 _____

Stakeholder N _____

Program Stakeholder Engagement Plan

Having identified a list of stakeholders during stakeholder identification and set up a stakeholder register and a stakeholder inventory, it is now time to devote some thought to how you are going to deal with them. Do not forget that some stakeholders could be less than friendly just because you ignored or did not involve them. A plan to engage these diverse stakeholders is essential and is done in the program stakeholder engagement plan (5.2) section in the stakeholder domain in the *Standard for Program Management*—Third Edition (2013).

Program Stakeholder Engagement Plan Instructions

The program stakeholder engagement plan includes the following:

Purpose: A brief introductory statement defining the purpose of the plan, such as:

> The program stakeholder engagement plan defines the program's strategy for effective stakeholder engagement throughout the program. Also, it ensures their commitment and support of the program. It is essential to understand and address the stakeholders' expectations and concerns to leverage them to support the program's business case and its continuation throughout its life cycle.

> The program stakeholder engagement plan builds on the organization's strategic plan, the business case, and the program charter. The stakeholder register is analyzed in consideration of these documents. This plan serves to ensure that stakeholder commitment to the program remains strong.

> This plan should identify guidelines for effective stakeholder engagement, provide insight as to how to best engage stakeholders especially to address any negative attitudes that any stakeholders may have about the program, and include metrics to show the effectiveness of the stakeholder engagement process. It considers the program's effect on the organization's culture and current major issues, and its likelihood to meet resistance and barriers to change. It assesses attitudes about the program, expectations about benefit delivery, the degree of support or opposition regarding benefits delivery, and the stakeholder's ability to influence the program's outcome. This plan is needed because of the complexities of many programs, the changes to be realized by them, and the interdependencies among program components and possibly other programs and projects under way in the organization. The plan supports the program's communications management plan and ongoing alignment of the program to the organization's strategic objectives.

> Since this plan relates to the program communications management plan, collectively these plans facilitate a clear understanding of issues, explain

program goals and objectives, and target the delivery of key messages and the engagement of key stakeholders when required. It is used to engage stakeholders (5.3).

It is an iterative document that should be reviewed regularly by the program management team and the Governance Board as new stakeholders become involved in the program at different phases of the life cycle or as other stakeholders have new interests in the program.

Stakeholder engagement importance: Stakeholder engagement is a key to successful program management. The program management team must work to ensure all stakeholders are engaged in the program and are supportive of it. This section describes how the program management team will ensure stakeholders are actively supporting the program so that any possible negative impacts can be easily resolved.

Stakeholder mitigation strategies: When stakeholders do have issues concerning the program, strategies are required to mitigate these negative impacts. This section describes the strategies the program management team will use to work with these stakeholders. For example, since programs represent organizational change, it may be appropriate for the program management team to conduct training so affected stakeholders understand why the program is being conducted so they can best adjust to the resulting changes from it. It also may be appropriate to prepare job aids for use once the program is transferred to ongoing operations to best handle the impact of the program on stakeholders. These job aids could include specific descriptions of processes to follow, guidelines to consider, and quick reference guides that are readily available.

Stakeholder commitment: This section describes the approaches the program management team will use to ensure that the stakeholders remain committed to the program. It includes metrics to gauge the level of stakeholder participation, the positive contributions stakeholders make to the attainment of the program's benefits, and the frequency of communications with the program manager and the team as well as metrics to assess the effectiveness of overall stakeholder engagement by the program manager and the team.

Stakeholder feedback: The program management team also must ensure that stakeholders have embraced the changes and that the mitigation strategies that have been prepared are effective. This section describes how the program management team will obtain feedback from stakeholders concerning these mitigation strategies to see if changes are required. It also will state the need for change orders as appropriate, whether the changes will result in the need to update any other program documents, and specific responsibility to implement the changes as necessary.

Approvals: This section contains the written approval of the stakeholder engagement plan by the program sponsor, program manager, members of the Governance Board, and others as appropriate.

Stakeholder Engagement Plan Template

<Insert Program Name>
Stakeholder Engagement Plan

Program name:	
Program manager:	PM's email address here as a hyperlink
Program sponsor:	
Proposed start date:	
Proposed end date:	
Prepared by:	
Program number:	
Revision history:	
Business unit:	

A. PURPOSE

A brief introductory statement defining the purpose of the stakeholder engagement plan, such as:

> The stakeholder engagement plan defines the program's strategy for effective stakeholder engagement throughout the program. It contains guidelines to ensure their commitment and support of the program.

_____

_____

_____

B. STAKEHOLDER ENGAGEMENT IMPORTANCE

This section describes how the program management team will ensure stakeholders are actively supporting the program so that any possible negative impacts can be easily resolved.

_____

_____

_____

C. STAKEHOLDER MITIGATION STRATEGIES

This section describes the strategies the program management team will use to minimize any negative impacts when stakeholders have issues with the program.

_____

_____

_____

D. STAKEHOLDER COMMITMENT

This section describes the approaches the program management team will use to ensure the stakeholders remain committed to the program and establishes metrics to use to ascertain stakeholder participation and the effectiveness of the program manager and the team in overall stakeholder engagement.

_____

_____

E. STAKEHOLDER FEEDBACK

This section describes how the program management team will obtain feedback from stakeholders concerning the mitigation strategies to see if changes are required.

_____

_____

F. APPROVALS

This section contains the approval of the stakeholder engagement plan by the program manager, program sponsor, program management office director, members of the Governance Board, and others as required.

SIGNATURES AND DATE APPROVAL OBTAINED

Program manager   _____

Program sponsor   _____

Program management office director   _____

Governance Board chairperson   _____

Governance Board member 1   _____

Governance Board member 2   _____

Governance Board member N   _____

Stakeholder 1   _____

Stakeholder 2   _____

Stakeholder N   _____

Component Stakeholder Engagement Guidelines

Since programs consist of projects and non-project work, these components also have stakeholders and interface with them. They also will likely interface with stakeholders at the program level. The program can help the components by suggesting sources of information such as the *Guide to the Project Management Body of Knowledge (PMBOK Guide)*—Fifth Edition (2013), providing program-unique guidance, and defining appropriate contact boundaries. These guidelines are noted in 5.2, Stakeholder Engagement Plan, in the *Standard for Program Management*—Third Edition (2013).

Component Stakeholder Engagement Guidelines Instructions

Purpose: Stakeholder engagement is equally as important at the component (project and non-project work) level as it is at the program level.

Each component will have stakeholders who will impact and influence the component. One recommended practice is for the program management team to issue stakeholder engagement guidelines to its components.

This document presents some guidelines for consideration as to how each component should effectively engage its stakeholders for component and program success. These guidelines should be reviewed periodically and updated as required:

1. As each component begins, the component manager and his or her team should identify relevant stakeholders and determine their level of interest in the component and influence over it. They may wish to interview stakeholders or to survey them as part of the identification process. They should document the results of their stakeholder identification in a register, similar to that at the program level.
2. The component managers should maintain this register of these stakeholders and should meet with them periodically to address their issues and concerns.
3. The component manager should ensure that the stakeholders receive the information they need regarding the status of the component, as stated in the component's program communications management plan. Information should be distributed according to the information distribution process.
4. The component managers should ensure that the information provided to the stakeholders meets their needs and should request feedback from them. Stakeholders, for example, may request other information about project performance, and this request may show a need to more actively engage the stakeholder and to update the information that is provided.
5. Component managers should work with the program management team to prepare the program stakeholder engagement plan. A joint planning session also is recommended as stakeholders involved with one component

may also be involved with other components, and these stakeholders then should receive higher priority at the program level in the program stakeholder engagement plan and the stakeholder inventory.

6. Component managers should inform the program manager about any issues they have with their stakeholders, as these issues may affect other components or the entire program.

7. If the component manager cannot resolve the stakeholder issues, he or she should ask the program manager for assistance and resolve the issues in a timely way.

8. Periodically, component managers should meet as a group to review these guidelines with the program manager and his or her team.

9. The guidelines also should be reviewed when the work of the component is complete and is transitioned to an operational unit or to the customer, or if the work of the component is terminated for any reason.

10. When new components are added to the program, the program management team should meet with the component manager to review these guidelines, explain their purpose, and solicit any suggestions for improvement.

11. Based on these feedback sessions, it may be necessary to revise these guidelines. If so, a member of the program management team should be assigned to implement these changes and inform the component managers of the updated guidelines.

12. The guidelines should be accessible by any member of the team.

Program Stakeholder Inquiry Register

In (5.3) the *Standard for Program Management—*Third Edition (2013) introduces an interesting type of program management detail, mentioning the natural curiosity of stakeholders and resultant questions. These questions generally would not be classified as issues, so they must be handled in some way. The manner in which they are subsequently addressed (and not forgotten) could be in the form of an issue or even a risk, and the result could have an impact on various plans, including but not limited to, the communications management plan.

While these inquiries could be placed into an issue register, one way to capture them would be in a stakeholder inquiry register. In any event, the description of such a registry below illustrates how these requests may be handled.

Program Stakeholder Inquiry Register Instructions

The program stakeholder inquiry register includes the following:

Purpose: A brief introductory statement defining the purpose of the program stakeholder inquiry register, such as:

This register captures gaps in the understanding of stakeholders for resolution. These gaps are determined when stakeholders ask questions through any communication channel and can be addressed through updates to the communications plan or the contents of the communications it prescribes. Register entries may result in generation any combination of issues, risks, or change requests.

1. **Inquiry identification number:** Assign a number to the customer inquiry.
2. **Date:** State the date on which the inquiry was entered.
3. **Originator:** Enter the name and contact information of the person entering the item into the registry. If this is not the person who received the request, include that person's name and contact information as well.
4. **Inquirer:** Enter the name and contact information of the stakeholder or other entity posing the inquiry.
5. **Owner:** Record the name and contact information for the program person designated to handle the inquiry to resolution.
6. **Status:** Record the status as either: Open, Owner Assigned, or Closed.
7. **Inquiry description:** Describe the inquiry, including, if appropriate, the reason why this additional communication was necessary.
8. **Impact:** Briefly describe the impact of the inquiry on cost, schedule, scope, benefits, documentation, quality, and any other significant aspects of this and other programs.
9. **Proposed resolution:** Enter the actions proposed to address the inquiry and, where appropriate, address the root cause of the inquiry.
10. **Resolution action notes:** Record notes documenting the final resolution of the inquiry, such as issue number generated, specific report modified, etc.
11. **Notes:** Use this field for any additional notes about the inquiry.

Stakeholder Impact and Issue Tracking Prioritization

Throughout the program with the numerous stakeholders involved Section 5.3 of *The Standard for Program Management*—Third Edition (2013) notes that an impact and issue tracking and prioritization tool is useful as there will be issues, which the program team will need to track to closure. We have prepared a template that you can use and tailor as appropriate as a Program Issue Register described in Chapter 7C in this book.

Program Communications Management Plan

Lots of internal and external stakeholders will be interested in what you are doing and how the program is progressing. Communications management and stakeholder engagement are closely related and are ongoing throughout the program as noted in The *Standard for Program Management*—Third Edition (2013). The Standard points out the importance of communications planning in Section 8.1.1 and is not kidding when it says communications planning is vital to program success. You need to decide how much of which information is sent to whom, how, and how often. For example, if you give just a little more information than a particular stakeholder needs—airing issues that you know will ultimately be resolved—you could wind up spending too much of your time answering questions about matters some stakeholders do not need to worry about yet. On the other hand, transparency demands that bad news gets to the right people in a timely manner. A balance is in order, and it is part of the art of program management. For example, failure to have a publishing schedule will mean missed deadlines for reports. Your program communications management plan addresses these issues and more.

Program Communications Management Plan Instructions

The program communications management plan includes the following:

Purpose: A brief introductory statement defining the purpose of the program communications management plan, such as:

> The program communications management plan describes the process of determining the information and communications needs of the program stakeholders and states which stakeholders need what information, to what granularity, when they need it, how it will be given to them, and by whom.

> This plan builds on the stakeholder engagement plan, which is developed earlier in the program. In fact, once the program communications management plan is approved, the stakeholder engagement plan can become an appendix to it, and the stakeholder register can be reviewed for needed updates (5.1). Effective communications are essential for program success. Program communications create a bridge between the various stakeholders in the program, connecting various cultural and organizational backgrounds, different levels of expertise, and various perspectives and interest in program execution. Although the plan is drafted early in the program, it is an iterative document that should be further refined as the program moves through the phases of the life cycle, especially since its development may entail the need to create additional deliverables. If the latter is the case, then the program work breakdown structure, program schedule, and program budget also will require updates.

Stakeholder communications requirements: This section builds on the stakeholder engagement plan as it describes the communications requirements of the program stakeholders. These requirements are defined by combining the type and format of the information needed with an assessment of the value of the information. It is important to ensure that program resources are expended only on communications that lead to program success, or where a lack of information may lead to program failure.

Information to be communicated: This section describes the information to be communicated, including language, format, content, and level of detail. As with many plan elements, this information will not be completely known at the outset. While all stakeholders require information about the program, the information needs of stakeholders vary greatly and may change during various program phases in the life cycle. To be effective, the information must be communicated in the right format, at the right time, and with the right impact. Efficiency also is important, as it focuses on providing only the information that is needed.

Program team member responsibilities: This section states which program team member is primarily responsible for the communications requirements of each of the program stakeholders.

People to receive the information: This section describes the specific people or organizational units that will receive each type of information that will be communicated. This section may change based on the stage of the process in the program life cycle, as different groups will have different levels of interest. Not everyone needs to communicate with everyone else; however, people do require information based on their roles and responsibilities in the program and their specific areas of interest.

Methods or technologies to use to convey the information: This section describes the methods to use to transfer information to the program stakeholders. Methods range from brief conversations to extended meetings or from one-page reports of progress to detailed status reports. To prepare this section consider factors such as the urgency of the need for the information, the available technology, the expected program staffing, the program duration, and the program environment.

Frequency of information to be provided: This section describes the frequency of information to be provided to each of the program stakeholders. Some stakeholders will require information on a daily basis, while others may only need program information before a stage gate review of the Governance Board.

Escalation process for communications issues: This section describes how communications issues will be escalated from project managers or program team members to the program manager or from the program manager to the sponsor or to the Governance Board. It should include time frames to resolve any issues at each level.

Methods to update the program communications management plan: This section describes the specific methods to use to update the program communications management plan as the program ensues. It is important to ensure that stakeholders receive the required information. If a number of issues surface that require escalation, this can serve as a trigger to update the plan. Similarly, if new stakeholders have an interest in the program, it may be appropriate to update the plan.

Glossary of common terminology: Each program has certain terms that are used by the various program stakeholders. Similarly, standard program management terms also are used. The same acronym may have vastly different meanings in different organizations. This section contains a glossary of common terms that stakeholders can use who are interested in the program to ensure there are no misunderstandings that can lead to communications breakdowns.

Guidelines for meetings and e-mail: Meetings and e-mail are common methods to distribute program information. This section describes how meetings are to be held: whether they are formal or informal, or collocated or e-meetings. It states the need for an agenda for each meeting, for minutes to be taken and distributed, and for action items or issues to be recorded and reviewed at subsequent meetings. This section also describes guidelines for the use of e-mail. It is important to ensure that guidelines are set as to how to identify critical program information, a specific timetable as to when a response is required, when an e-mail should be sent to the entire program team or to only key team members, manual or automated archiving processes, and the need for confidentiality and security requirements for certain information in e-mails.

Approvals: This section contains the written approval of the program communications management plan by the program sponsor, program manager, program management office director, members of Governance Board, and other stakeholders.

Program Communications Management Plan Template

<Insert Program Name>
Program Communications Management Plan

Program name:	
Program manager:	PM's email address here as a hyperlink
Program sponsor:	
Actual start date:	
Approved end date:	
Program number:	
Revision history:	
Business unit:	

A. PURPOSE

A brief introductory statement defining the purpose of the program communications management plan, such as:

> The program communications management plan describes the process of determining the information and communications needs of the program stakeholders and states which stakeholders need what information, when they need it, how it will be given to them, and by whom.

_____

_____

_____

B. STAKEHOLDER COMMUNICATIONS REQUIREMENTS

This section describes the communications requirements of the program stakeholders. They are defined by combining the type and format of the information needed with an assessment of the value of the information.

_____

_____

_____

C. INFORMATION TO BE COMMUNICATED

This section describes the information to be communicated, influencing language, format, content, and level of detail. The information needs of program stakeholders vary greatly and may change during the various program phases in the life cycle.

_____

_____

_____

D. PROGRAM TEAM MEMBER RESPONSIBILITIES

This section states which program team member is responsible for the communications requirements of each program stakeholder.

_____

_____

_____

E. PEOPLE TO RECEIVE THE INFORMATION

This section describes the specific people or organizational units that will receive each type of information that will be communicated. People require information based on their roles and responsibilities in the program and their specific areas of interest.

_____

_____

_____

F. METHODS OR TECHNOLOGIES TO USE TO CONVEY THE INFORMATION

This section describes the methods to use to transfer information to the program stakeholders. Consider items such as the urgency of the need for the information, the available technology, expected program staffing, program duration, and the program environment.

_____

_____

_____

G. FREQUENCY OF INFORMATION TO BE PROVIDED

This section describes the frequency of information to be provided to each of the program stakeholders, as some may require only infrequent communications, and others may require information on a daily basis.

_____

_____

_____

H. ESCALATION PROCESS FOR COMMUNICATIONS ISSUES

This section describes how communications issues will be escalated from project managers or project team members to the program manager, and from the program manager to the program sponsor or to the Governance Board. It should include time frames to resolve any issue at each level.

_____

_____

_____

I. METHODS TO USE TO UPDATE THE PROGRAM COMMUNICATIONS MANAGEMENT PLAN

This section describes the specific methods to use to update the program communications management plan as the program ensues. It is important to ensure stakeholders receive the required information, and stakeholders will change over the life of the program.

_____

_____

_____

J. GLOSSARY OF COMMON TERMINOLOGY

This section contains a glossary of common terms that are used by the program stakeholders and standard program management terms to ensure there are no misunderstandings that can lead to communications barriers.

_____

_____

_____

K. GUIDELINES FOR MEETINGS AND E-MAIL

This section describes how meetings are to be held and their specific format. It states the need for an agenda for each meeting, for minutes to be taken and distributed, and for action items or issues to be recorded and reviewed at subsequent meetings. It also describes guidelines for the use of e-mail to state critical program information, a specific timetable in which a response is required, when e-mail should be sent to the entire team or to only key team members, and the need for confidentiality and security requirements for certain information in e-mails.

_____

_____

_____

L. APPROVALS

This section contains the approval of the program communications management plan by the program sponsor, program manager, program management office director, members of the Governance Board, and other key stakeholders.

SIGNATURES AND DATE APPROVAL OBTAINED

Program manager _____

Program sponsor _____

Program management office director   _____

Governance Board chairperson   _____

Governance Board member 1   _____

Governance Board member 2   _____

Governance Board member N   _____

Stakeholder 1   _____

Stakeholder 2   _____

Stakeholder N   _____

Communications Log/Stakeholder Engagement Log

Your program cannot be sure of consistent and uniform information without a communications log or a stakeholder engagement log. It consists of meeting minutes, e-mails, reports and memos, presentations, etc. It is usually implemented in a Web portal or perhaps a database.

Communications Log/Stakeholder Engagement Log Instructions

Purpose: The communications log or stakeholder engagement log is used throughout the program to document program communications to the various program stakeholders. It also is used to show other communication items that were not in the program communications management plan but were requested by stakeholders, and how the stakeholders received the needed information. These data then are used to determine whether changes to the program communications management plan and/or the stakeholder engagement plan are warranted. This log is prepared by a program team member or the program manager.

This communications log complements the program communications management plan.

The communications log is designed as a table, and a description of its contents follows:

1. **Communications requirement:** List the stakeholder communications requirement as described in the program communications management plan.
2. **Type of communication:** State the type of communication. Examples include upcoming meetings, meeting agendas, status reports, memos, press releases, risk register, benefit register, program schedule, financial reports, meeting minutes, formal reports, lessons learned, newsletters, etc.
3. **Frequency:** State how often the specific type of communication is to be prepared. Examples include daily, weekly, monthly, quarterly, at the end of a phase gate or program review, at the end of an audit, according to

schedule milestones, when deliverables are completed, at the end of the program, etc.

4. **Prepared by:** List the person who prepared the communications item and his or her contact information.
5. **Delivered to:** List the people who received the communications item and their contact information.
6. **Delivery method:** State how the communication was delivered, e.g., e-mail, posted in a discussion forum or in a blog, hard copy, newsletter, demonstrations, brochures, etc.
7. **Date:** State the date the communications item was delivered.
8. **Feedback measures:** State the approach used by the program management team to determine if the communications item was considered effective by the recipients, such as an interview, phone call, e-mail, or survey. Use this column to determine whether the communications item was considered effective and addressed the stakeholder's requirements to see if any changes are needed.
9. **Communications request:** Describe any requests for unscheduled communications by program stakeholders.
10. **Requested by:** List who requested the communications item and his or her contact details.
11. **Prepared by:** State who prepared the needed information and his or her contact details.
12. **Recipients:** State who received the information and his or her contact details.
13. **Delivery method:** State how the information was delivered.
14. **Date:** State the date the communications item was delivered.
15. **Notes:** Use this field for any additional notes about the communications item.

Communications Strategy

According to the *Standard for Program Management*—Third Edition (2013), in section 8.1.2.1, being an effective communicator is a core competency for program managers. Since the program manager is the key communicator, the Standard suggests that he or she develop a communications strategy, which is used throughout the program. It notes it may be only a 'quick reference' for the program manager in order that each stakeholder receives the information he or she needs about the program. This strategy builds on the stakeholder register, the stakeholder engagement log or communications log, and the program communications management plan. It is updated regularly as stakeholders and messages change

Communications Strategy Instructions

The communications strategy includes the following:

Purpose: A brief introductory statement defining the purpose of the communications strategy, such as:

> The communications strategy ensures that stakeholders receive the information they need about the program.

Communications are especially important on programs given the numerous stakeholders involved, the high level of interest in the outcomes of the program in terms of achieving benefits, and the link between the program's objectives and the organization's strategic objectives.

It is suggested in section 8.1.2.1 as a reference guide for the program manager in program communications considerations in the *Standard for Program Management*—Third Edition (2013), but it is an iterative document and should be reviewed on a periodic basis to ensure that stakeholders receive the information they need in a timely way and that the information provided is relevant and accurate.

Organization's communications strategy: If the organization has a communications strategy, it should be noted in this section. The section also should state whether this strategy is to be used without change or whether tailoring is needed to meet the specific program requirements. If this strategy is to be used without any changes, there is no need to complete the rest of this template except for the approvals section.

Constraints: This section documents those items that may limit the program team's communications options. Examples are stakeholders who are located in different geographic areas to ensure they are not affected adversely when virtual meetings are held; the need to use a common language for communications, such the standard English vocabulary of 4,000 words; an inability to access common systems because of the lack of software compatibility; and the use of different technical capabilities.

Assumptions: This section describes those items that are considered to be true, real, or certain concerning the communications strategy but for which adequate validation is not available. Examples may be regulatory requirements, changes in technology, the need to consult with advisory boards, media involvement at certain phases of the program, involvement of consumer groups, terms and conditions and other confidential clauses in contracts, and organizational requirements.

Information transfer from components to the program: This section states the format and process the components will use to transfer information from the component to the program and how often information will be transferred. It also describes the mechanisms to ensure that the information that is transferred is timely and accurate. To ensure stakeholders receive needed information, major component milestones should be addressed as part of the program's master schedule, and information should be provided when these milestones are met.

Assessment of potential communications channels: This section states the potential communications channels in the program given the number of stakeholders. The program manager then can determine who needs to communicate with whom and how often communications should occur in order that there is an optimal exchange of information with stakeholders.

Resolution of stakeholder issues and concerns: This section describes the methods to be used to resolve any stakeholder issues and concerns. It describes how component managers will escalate issues to the program manager. It states the items the program manager can resolve on his or her own and the process to use to escalate items to the Governance Board for resolution.

Approvals: This section contains the written approval of the communications strategy by the program sponsor, program manager, program management office, members of the Governance Board, and others as appropriate.

Communications Strategy Template

<Insert Program Name>
Communications Strategy

Program name:	
Program manager:	PM's email address here as a hyperlink
Program sponsor:	
Proposed start date:	
Proposed end date:	
Prepared by:	
Program number:	
Revision history:	
Business unit:	

A. PURPOSE

A brief introductory statement defining the purpose of the communications strategy, such as:

The communications strategy ensures that stakeholders receive the information they need about the program.

_____

_____

_____

B. ORGANIZATION'S COMMUNICATIONS STRATEGY

This section states whether the organization's communications strategy will be followed in the program.

_____

_____

_____

C. CONSTRAINTS

This section documents those items that may limit the program team's communications options.

_____

_____

_____

D. ASSUMPTIONS

This section describes those items that are considered true, real, or certain concerning the communications strategy.

_____

_____

_____

E. INFORMATION TRANSFER FROM COMPONENTS TO THE PROGRAM

This section states the format and process the components will use to transfer information from the component to the program and how often information will be transferred. It describes mechanisms to ensure the information that is transferred is timely and accurate.

_____

_____

_____

F. ASSESSMENT OF POTENTIAL COMMUNICATIONS CHANNELS

This section states the potential communications channels in the program given the number of stakeholders.

_____

_____

_____

G. RESOLUTION OF STAKEHOLDER ISSUES AND CONCERNS

This section describes the methods to be used to resolve any stakeholder issues and concerns.

_____

_____

_____

H. APPROVALS

This section contains the approval of the communications strategy by the program manager, program sponsor, program management office director, members of the Governance Board, and others as required.

SIGNATURES AND DATE APPROVAL OBTAINED

Program manager _____

Program sponsor _____

Program management office director _____

Governance Board chairperson _____

Governance Board member 1 _____

Governance Board member 2 _____

Governance Board member N _____

Stakeholder 1 _____

Stakeholder 2 _____

Stakeholder N _____

Chapter 6

Governance

Effort and courage are not enough without purpose and direction.

—John F. Kennedy

Since the 1980s, there has been increased attention in organizations of various types as to the importance of programs and projects as strategic assets. Increasingly, the management by projects/programs is a reality,

In programs because of their longer duration and the projects and non-project work that comprise them, a Governance Board (or a Steering Committee or Program Board) typically is used to ensure the program continues to support the organization's strategic objectives and to approve the program to move to a new stage gate when appropriate. The Governance Board also can conduct periodic performance reviews and serve as a forum for the program manager to raise issues and risks for resolution when he or she feels that they require additional attention or if they may affect other program or project work elsewhere in the organization.

As the program is approved based on its business case, a Governance Board, often a Portfolio Review Board, makes the decision that it should be part of the portfolio. In most organizations, a different Governance Board then works with the program as it moves throughout the life cycle. Some programs may have multiple Governance Boards for complex programs, ones that are part of a consortium, or ones that may have government and private sector organizations. On some programs, the program manager and his or her team serve as a Governance Board for the projects and non-project work of the program; however, the best practice is for the program to have a single Governance Board as recommended in the *Standard for Program Management*—Third Edition (2013).

A high-level description of the tasks in the *Examination Content Outline* (PMI, 2011) for the Governance domain follows:

- Develop program and project management standards and structure using best practices to promote efficiency and consistency among projects and program objectives
- Select a governance model to deliver program objectives in alignment with organizational governance requirements
- Obtain authorization and approvals through stage gate reviews
- Evaluate key performance indicators to monitor benefits throughout the life cycle
- Develop and use the program management information system and integrate processes to manage program information and communicate status to stakeholders
- Evaluate new and existing risks that impact strategic objectives
- Establish escalation processes to ensure risks are handled at the appropriate level
- Develop and contribute to an information repository with lessons learned from programs
- Identify and apply lessons learned to foster future program or organizational improvement
- Monitor the business environment to ensure the program remains aligned with the organization's strategic objectives
- Develop and support the program integration management plan to further support program strategic objectives

This list shows the importance of governance as a major theme in program management.

The *Standard for Program Management*—Third Edition (2013) focuses as well on the importance of governance as a domain emphasizing the importance of a governance structure to best achieve the program's planned benefits. It shows the relationship of governance to benefit management and the interaction as well with stakeholder engagement, since members of the Governance Board are key stakeholders. It further states that in terms of program management success by incorporating it into business processes requires a strong emphasis on governance among other items. Therefore, processes and procedures for program management oversight and decision making are required.

Suggested artifacts for this domain follow.

Program Governance Plan

As discussed, while this topic may seem to be somewhat mundane, at first glance, it is a prominent area of current interest to ensure programs remain in alignment with the organization's strategic objectives. Entire books are devoted to various aspects of governance.

When some aspect of the program threatens to get out of control, action is required. Resulting inter-organizational and intra-organizational struggles can be divisive and counterproductive. Who has the authority to make such decisions, and how will they keep tabs on things? Of course, at times, the program manager may be too close to the situation to know what should be done. The governance process is the answer. Thus, the importance of preparing governance plan (6.2.4) and establishing a program governance structure process (6.2.4.2) in the *Standard for Program Management*—Third Edition (2013) is clear.

Program Governance Plan Instructions

The program governance plan includes the following:

Purpose: A brief introductory statement defining the purpose of the program governance plan, such as:

> The governance plan describes the process that will be followed to execute the program's governance activities. Its focus is on goals, structure, roles and responsibilities, and overall logistics for the Governance Board. It serves to ensure the program's goals remain in alignment with the strategic goals of the organization and that its proposed benefits will be met.

> The program governance plan is a subsidiary plan to the program management plan. Effective program governance is key to successful program management. The program Governance Board is a major stakeholder in every program, to focus on benefits realization, stage gate reviews, and effective decision making. The Governance Board, also known as a steering committee, oversight committee, or the Board of Directors, in many organizations, emphasizes proactive actions rather than the need for corrective actions. The Governance Board, as noted through this plan, serves to identify, analyze, and respond to internal and external events in the program and to change, initiate, terminate, or transition components as required. The governance plan states the process to ensure that decisions are made in a timely manner to not impede program progress.

> This plan is an iterative document and should be reviewed periodically by the program management team and the members of the Governance Board, as the work of the program continues throughout the various phases in its life cycle.

> In the Project Management Institute's *Standard for Program Management*—Third Edition (2013), the Governance Board may be the approval authority for the business case (3.1.1) and the program charter (8.3.1.6).

Goals: This section describes the goals for program governance, such as the importance of ensuring that the program remains in alignment with the organization's strategic goals and objectives. It lists the goals for each of the program's components as well as the goals for the overall program and how the benefits will be delivered. It describes how progress toward these goals will be communicated. Additionally, it ensures that interfaces between the program with other programs, projects, and ongoing operations within the organization are managed effectively to minimize conflicts and maximize opportunities. Furthermore, this section discusses the role of risk management in governance activities. Recognizing that often programs involve alliances with other organizations, this section describes the adherence to key policies, procedures, and standards as applicable.

Organizational structure: This section describes the structure of the Governance Board. Existing organizational charts are used since they show accountabilities and authority levels within the organization. Also, different programs in the organization will use different structures, meaning that the emphasis in this section is to determine a structure that will facilitate effective program governance.

Roles and responsibilities: This section lists the members of the Governance Board and the Board's specific responsibilities. It should be noted that while consensus is desired among the Board members, it is not a prerequisite. An executive director or a program sponsor typically chairs the Board and is the ultimate decision maker. This individual generally is a senior manager and provides organizational resources to the program. He or she has overall responsibility for program success. If partner organizations are part of the program, they should be represented on the Governance Board. On certain programs, the customer may be represented. This section states the involvement of the portfolio manager, business change manager, functional managers, program sponsor, program manager, director of the program management office, project managers, and program team members as appropriate in terms of serving as Board members or in interactions with the Board. This section also describes specific accountabilities for benefits realization, communication with stakeholders, and oversight of the program and its components.

Governance decisions: This section describes the decision-making approach the Board will follow. It states how decisions will be documented and communicated to program stakeholders, perhaps through the use of a governance decision register. It also describes an escalation process to follow if the Board does not feel it is empowered to make certain types of decisions. Since decisions of the Board impact the overall benefits delivery of the program and strategic alignment, each member of the Board should be aware of the organization's strategic plan, vision, mission, and values.

Meeting schedule: This section presents an overview of the frequency of the Governance Board's meetings. Regularly scheduled meetings should be shown on the program's schedule. It also notes that meetings may be called as needed and criteria to consider as to when to call an ad hoc meeting. It describes the process to be used to conduct each meeting, including responsibilities for handling logistics, preparing and distributing the agenda, taking and distributing meetings, escalating issues as required and documenting decisions that are made. It describes others who can attend various meetings. It lists phase gate meetings and program 'health' checks or periodic progress review meetings.

Gate review requirements: Since programs typically span a longer duration than projects and are more complex, phase-gate reviews are a recommended best practice, especially when programs have completed a phase in the life cycle. This section states the requirements for these reviews. They serve as go/no-go decisions. These gate reviews assist in overall program monitoring and control to ensure that not only is the program being managed effectively but also program components are being managed as planned. These reviews also provide a way to assess the strategic and quality criteria, such as ensuring that the program remains aligned with the organization's overall strategic direction; that program benefits are being realized as planned; that the level of risk to the program is one that fits the organization's overall tolerance for risks; that variances in scope, schedule, and cost are consistent with the organization's practices; and that best practices in program management continue to be followed. This section discusses what will occur at these reviews, when they will be scheduled, the items to be covered, roles and responsibilities, and measurement criteria.

Program performance review requirements: In addition to gate reviews, the Governance Board typically reviews the program at various times. These reviews concentrate on overall program performance and management and to ensure planned benefits will be realized and ultimately sustained. They evaluate existing plans to see if they are effective or require changes; the program tools and techniques that are being used, such as the program management information system; overall performance against desired outcomes and benefits realization; alternatives to contribute to overall program success; and whether the existing processes and procedures remain helpful or need change.

Process to initiate, terminate, and transition components: The Governance Board approves the initiation of new components to be part of the program and determines whether a component should be terminated because it has completed its deliverables and delivered its expected benefits or should be terminated because it no longer is supportive of the program's goals and objectives. It also approves when a component should be transitioned to ongoing operations, such as to a product or customer support group, or to the customer when it has completed its goals and objectives and its benefits have been realized. This section describes the process the Board follows for initiation, termination, and transition, and the criteria it uses.

Process to close the program: Additionally the Governance Board makes the recommendation regarding program closure. This section describes the process that is followed to ensure the recommendations for closure support the organization's strategic goals and objectives.

Issue escalation process: On programs, project and operations managers may escalate issues to the program manager for resolution, and the program manager in turn may escalate them the Governance Board. This section descries the issue escalation process to be followed at all levels and notes when key stakeholders need to be part of it.

Program success criteria: Different success criteria are appropriate for each program. This section states the minimum success criteria for the program and the metrics to be used to assess progress.

Program approach and plans: The Governance Board is responsible for approving the overall program approach and also for approving the approach each component will use. This section states the process to be followed and also the approach to be used to manage and control the program's components.

Assessment of effectiveness: On a periodic basis, it is incumbent to ensure that the program governance plan is effective and to determine whether any changes are required to it. It also is necessary to see whether or not the principles set forth in this plan are being followed. This section describes how effectiveness is determined and who is responsible for this evaluation.

Approvals: This section contains the written approval of the governance management plan by its members and any other key stakeholders as appropriate.

Program Governance Plan Template

<Insert Program Name>
Governance Management Plan

Program name:	
Program manager:	PM's email address here as a hyperlink
Program sponsor:	
Actual start date:	
Approved end date:	
Program number:	
Revision history:	
Business unit:	

A. PURPOSE

A brief introductory statement defining the purpose of the program governance plan, such as:

> The program governance plan describes the process that will be followed to execute the program's governance activities. Its focus is on goals, structure, roles and responsibilities, and overall logistics of the Governance Board. It serves to ensure the program's goals remain in alignment and that the program's proposed benefits will be met.

_____

_____

_____

B. GOALS

This section describes goals of program governance in the program such as to ensure that the program remains aligned with the organization's strategic goals and objectives and that interfaces are managed effectively. It also discusses the role of risk management in governance activities and states the importance of adherence to key policies, procedures, and standards as applicable.

_____

_____

_____

C. ORGANIZATIONAL STRUCTURE

This section describes the structure of the Governance Board.

_____

_____

_____

D. ROLES AND RESPONSIBILITIES

This section lists the members of the Governance Board and specific responsibilities. It describes specific accountabilities for benefits realization, stakeholder communications, and oversight of the program and its components.

_____

_____

_____

E. GOVERNANCE DECISIONS

This section describes the decision-making approach the Board will follow. It states how decisions will be documented and communicated to stakeholders and describes an escalation process to follow if the Board does not feel it is empowered to make certain types of decisions.

_____

_____

_____

F. MEETING SCHEDULE

This section presents an overview of the frequency of Board meetings and notes that meetings may be called as needed. It describes the process to follow for meeting logistics and who can attend various meetings.

_____

_____

_____

G. GATE REVIEW REQUIREMENTS

This section states the requirements for program phase-gate reviews. It describes the purpose of these reviews and the items that are covered during each review.

_____

_____

_____

H. PROGRAM PERFORMANCE REVIEW REQUIREMENTS

This section describes the process to follow for the Board to review overall program performance at various times. It discusses the objectives of these reviews.

_____

_____

_____

I. PROCESS TO INITIATE, TERMINATE, AND TRANSITION COMPONENTS

This section describes the process and the criteria the Board follows to initiate new components to be part of the program, to terminate existing components, and to transition components when the work is complete and benefits have been realized.

_____

_____

_____

J. PROCESS TO CLOSE THE PROGRAM

This section describes the process and the criteria the Board follows to recommend program closure. It ensures the closure process is followed and that program closure is consistent with the organization's strategic goals and vision.

K. ISSUE ESCALATION PROCESS

This section describes the process to be followed on the program to escalate issues from component managers to the program manager and from the program manager to the Board. It describes when key stakeholders need to be engaged.

L. PROGRAM SUCCESS CRITERIA

This section describes the specific success criteria for the program and how it will be measured.

M. PROGRAM APPROACH AND PLANS

This section describes the overall approach the program and its components will follow to achieve their goals and the framework to be used to manage and monitor the components.

N. ASSESSMENT OF EFFECTIVENESS

This section describes how program governance is assessed for its effectiveness in terms of overall delivery of program benefits and describes who is responsible for this evaluation.

O. APPROVALS

This section contains the approval of the program governance plan by the members of the Governance Board and other key stakeholders as required.

SIGNATURES AND DATE APPROVAL OBTAINED

Governance Board chairperson _____

Governance Board member 1 _____

Governance Board member 2 _____

Governance Board member 3 _____

Stakeholder 1 _____

Stakeholder 2 _____

Stakeholder N _____

Change Management Plan

According to the *Standard for Program Management*—Third Edition (2013), program managers accept and embrace change—more so than project managers. Change is inevitable. Success requires control of change and, where that is not possible or desirable, adaptation to it. While this plan possibly could be included in any of the domains according to the *Standard for Program Management*—Third Edition (2013), we have elected to include it here since section 6.2.10 focuses on the importance of monitoring progress and the need for change. As noted, the Governance Board determines the types of changes a program manager should be able to decide alone and the types that should be escalated to the Board. The objective is to enable the Governance Board to provide support when changes are required. The goal is to position the program to best respond to any changes that occur at the organizational level so it can continue to support the organization's strategic goals and deliver its benefits. A change management plan is a guidebook to make this happen.

Change Management Plan Instructions

The change management plan will include the following:

Purpose: A brief introductory statement defining the purpose of the change management plan, such as:

The change management plan describes the processes and procedures to influence and manage program changes. It outlines the approaches to follow when there is a program change and assesses its impact on the overall program.

It also analyzes the impact of changes on the overall outcome of the program and assigns roles and responsibilities to ensure the impact of the change is assessed appropriately, especially since other programs may be affected. It focuses on factors that may influence program change to maximize them for the benefits of the program. It helps to coordinate changes across the program.

Although the plan is drafted early in the program, it is an iterative document that should be further refined as the work of the program continues throughout the various phases in its life cycle. It involves redirecting or modifying the program as required based on changes that do in fact occur. It identifies standard change process information used within the organization as well as program-specific information. It should be provided to all program stakeholders.

Definition of the change process: This section describes the change process to be followed in the program. It differentiates between program changes and product/services/results changes accordingly but integrates the product/services/results changes into the overall plan. It notes that changes may be either internal to the program or external to it. It identifies the key stakeholders who may be involved and the communication process to follow. It describes how the program manager must communicate the importance of changes to affected stakeholders. The program manager must plan for change, provide resources and support for changes, monitor and track changes, provide feedback on changes to stakeholders, and manage any issues stakeholders may have with each change. It states that changes may be approved, deferred, or rejected; the types of changes to be escalated to the Governance Board; the priority of the change; and its severity in terms of the program. It describes the specific roles and responsibilities of the program stakeholders, including Governance Board members, regarding change management.

Change type: This section shows the types of changes that may occur in the program. Some changes may be mandatory, while others are ones that may benefit the program or may clarify an existing requirement. The emphasis is on consideration of the implications of each change in terms of its impact on the program's goals or the organization's strategic goals. Regardless, different types of changes will then require different levels of detail to be provided concerning the change and will require different approval conditions. Some changes will involve more risks than others. The type of change also may impact other program documents that may require modification if the change is approved and implemented.

Identification of controlled/configuration items: This section describes the controlled items in the program, sometimes referred to as configuration items. The emphasis of the identification of the items to be controlled should focus on physical items, documents, forms, and records. They are items that may be legal and regulatory requirements, health and safety guidelines, ones that are sent to contractors, ones that have an impact on the process or deliverables that are outside of the current program, and ones that involve the program's deliverables. For example, certain program artifacts, such as the program charter, program work breakdown structure, program management plan, program governance plan, benefits realization plan, stakeholder engagement plan, approved schedule, approved budget, risk register, metrics, and contracts, are examples of possible controlled items or configuration items that require control. Each configuration item should be uniquely identified for control, processing, and tracking. Ideally, the configuration item should be linked to the program work breakdown structure program package identification number.

Change evaluation and approval processes and procedures: This section describes how a change request is to be used, how the change request will be analyzed, how decisions will be made based on the change impact analysis, and metrics to be reported as to the impact of the change. It notes the importance of documenting all change requests. It also describes how notifications to stakeholders regarding the disposition of the change will be made, states the process to implement the approved changes, and verifies that the implementation has occurred.

Change control: This section describes the process to be followed for change control. It explains how the change management process is integrated with the program management information system. It notes the importance of a full audit trail and reporting of the status of change requests. It describes how specific change control systems, such as those for contracts or cost, are part of an integrated change control system across the program.

Change status accounting and metrics: This section describes how information about the changes is communicated, and actions are validated. It serves as a feedback mechanism for the overall program change management process. It describes the process to follow to acquire and enter information about the change into the appropriate repository. It also provides the data through established metrics to validate if the approved changes have been consistent with the program's objectives and can be tracked to its scope and evolution. The status of approved changes is tracked to inform the program manager and other stakeholders about the results of an approved change. Metrics are part of the change management plan and include items such as what was changed, when it was changed, what impact it has on the previously captured data and potential effects of new data, and the impact on other related programs and projects. The purpose is to assist the program manager in identifying trends in the productivity of implementing various changes, determine if estimating techniques are still valid, track customer satisfaction, and track rework. The metrics should be ones that are analyzed and used to facilitate communication across the program. Ad hoc and periodic reports to be prepared for stakeholders are included in this section.

Change verification and audits: This verification process of this section ensures that the program change management goals are achieved through a systematic comparison of requirements with the initial, interim, and final results.

The audit process of this section ensures that program configuration items have been handled according to their defined documentation. The actual process used is compared to the documented process to uncover any deficiencies or areas for improvement. This section also describes how results from the audits will be communicated to program stakeholders.

Updates to other documents: This section describes the type of changes that may require the need to update program documents such as the governance decision register, the program plan, or the governance plan. It also notes changes that may have an impact on the program's budget and finances, organizational structure, or roles and responsibilities.

Approvals: This section contains the written approval of the change management plan by the program sponsor, program manager, program management office director, members of the Governance Board, and other stakeholders.

Change Management Plan Template

<Insert Program Name>
Change Management Plan

Program name:	
Program manager:	PM's email address here as a hyperlink
Program sponsor:	
Actual start date:	
Approved end date:	
Program number:	
Revision history:	
Business unit:	

A. PURPOSE

A brief introductory statement defining the purpose of the change management plan, such as:

The change management plan describes the processes and procedures to influence and manage program changes. It outlines the approaches to follow when there is a program change and assesses its impact on the overall program.

B. DEFINITION OF THE CHANGE PROCESS

This section defines the change process to be followed in the program. It notes the different changes that may affect the program, the key stakeholders and communications processes to follow, the disposition of changes, the priority of changes, and the roles and responsibilities of the stakeholders, including members of the Governance Board.

_____

_____

_____

C. CHANGE TYPE

This section shows the types of changes that may occur in the program. It explains that based on the types of changes, different levels of detail concerning the change will be required as well as different approval requirements. The section also states the effect of the type of change on the impact of other program documents if the change is approved and implemented. It notes the type of changes that require approval by the Governance Board.

_____

_____

_____

D. IDENTIFICATION OF CONTROLLED/CONFIGURATION ITEMS

This section describes the controlled items in the program or configuration items. Each one should be uniquely identified for control, processing, and tracking.

_____

_____

_____

E. CHANGE EVALUATION AND APPROVAL PROCESSES AND PROCEDURES

This section describes how a change request is to be used, how it will be analyzed, how decisions will be made based on the analysis, and how metrics will be reported. It also includes how notifications to stakeholders concerning the disposition of the change request will be made, the process to implement approved changes, and the process to verify that implementation has occurred.

_____

_____

_____

F. CHANGE CONTROL

This section describes the process to be followed for change control and explains how the change management process is integrated with the program management information system. It notes the importance of a full audit trail and reporting on the status of change requests.

G. CHANGE STATUS ACCOUNTING AND METRICS

This section describes how information about changes is communicated, and actions are validated. It describes the process to follow to acquire and enter information about the change into the appropriate repository and provides data through established metrics to validate if the approved changes are consistent with the program's objectives and can be tracked to its scope and evolution. The metrics to be collected are described in this section as well as various ad hoc and periodic reports to be prepared.

H. CHANGE VERIFICATION AND AUDITS

This verification part of this section states how the program change management goals have been achieved through a systematic comparison of requirements with the initial, interim, and final results. The audit part of this section states that the program configuration items have been handled as planned and compares the actual process used to the documented process. It also describes how results from the audits will be communicated to stakeholders.

I. UPDATES TO OTHER DOCUMENTS

This section describes other program documents that may require updates based on the type of change such as the governance decision register, governance plan, program management plan, and plans at the component level. It notes whether changes in budgeting and funding, the program's structure, and roles and responsibilities are needed.

J. APPROVALS

This section contains the approval of the change management plan by the program sponsor, program manager, program management office director, members of the Governance Board, and other key stakeholders.

SIGNATURES AND DATE APPROVAL OBTAINED

Program manager	_____
Program sponsor	_____
Program management office director	_____
Governance Board chairperson	_____
Governance Board Member 1	_____
Governance Board Member 2	_____
Governance Board Member N	_____
Stakeholder 1	_____
Stakeholder 2	_____
Stakeholder N	_____

Change Request Form

Change requests are inevitable, and there are a variety of possible change forms. The key here is that this form should be focused at the *program* level. Change requests are discussed in section 6.2.10 in the Governance domain in the *Standard for Program Management*—Third Edition (2013) and while mentioned in other parts of the document are included in this part of the book. As noted in the *Standard for Program Management*—Third Edition (2013), many change requests at the program level involve modifying the program's overall strategy, plan, or use of resources.

Change Request Form Instructions

Purpose: The change request form is used throughout the program. It is inevitable that programs will have changes throughout the program life cycle. The component projects also will have changes. Each project should manage its changes and use a change request form as well once a change is identified.

This change request form complements the program's change management plan and change request log.

The change request form is designed as a table, and a description of its contents follows:

1. **Change identification number:** Assign a change number to the change request.
2. **Date:** The date on which the change request was entered.
3. **Requested by:** List the person who requested the change and provide contact information.
4. **Change type:** Describe the type of change. Use the types in the change management plan, such as one involving technical aspects or the program's scope, schedule, cost, benefits, requirements or contracts. List changes such as ones involving a change to the program's strategy, plan, or use of resources.
5. **PWBS number(s):** Link the change to the corresponding program package in the program work breakdown structure.
6. **Change description:** Describe the change and why it is significant to the program.
7. **Impact:** Briefly describe the impact of the change on cost, schedule, scope, documentation, quality and any other significant aspects of this and other programs. If this is an involved subject, consider using the impact analysis format shown in Chapter 7D in this book.
8. **Notes:** Use this field for any additional notes about the change such as whether the changes are feasible, can be supported operationally, and their level of risk.

Change Request Log

A detailed list of all changes, whether approved, rejected, deferred, or modified, must be maintained. This log complements the change management plan and change request form and is included in this domain as it is stated in section 6.2.10 in the *Standard for Program Management*—Third Edition (2013).

Change Request Log Instructions

Purpose: The change request log is used throughout the program to record changes as they are identified, and to describe the status of the request.

It is inevitable that programs will have changes throughout the program life cycle. The component projects also will have changes. Each project should manage its changes and report them to the program management team, especially if it seems as if the project change will affect the program. Changes should be tracked until they are approved, deferred, or rejected. Once a decision is made concerning the change, then the person who requested the change should be notified. If the change is implemented, its status should continue to be tracked through this log until the implementation is complete.

It is important to ensure that the benefits to be realized through the program are not compromised by the change.

This change request log complements the program's change management plan. In the Project Management Institute's *Standard for Program Management—* Third Edition (2013), the change request log is noted in section 6.2.10. The change request log is designed as a table, and a description of its contents follows:

1. **Change identification number:** Assign a change number to the change request.
2. **PWBS number:** Link the change to the corresponding program package in the program work breakdown structure.
3. **Change description:** Describe the change and why it is significant to the program.
4. **Requested by:** List the person who requested the change and provide contact information.
5. **Date:** State the date the change was requested and added to the log.
6. **Change type:** Describe the type of change. Use the types identified in the change management plan.
7. **Priority/severity:** Describe the priority of the change request, such as emergency, urgent, high, medium, or low. Note the severity, such as catastrophic, major, minor, or an enhancement.
8. **Change impact:** State the impact of this change to the program. Note the configuration items identified in the change management plan that are affected.
9. **Cost:** Describe the expected cost of the change.
10. **Level of effort:** State the level of effort required to implement the change if it is approved. Note whether the program's schedule will require revision, the other program management processes that will require revision, and whether any new program components will be needed or existing components terminated.
11. **Feasibility:** Describe the feasibility of implementing the change, such as technical, environmental, economic, or ethical.
12. **Risk/opportunity:** Describe the risks associated with implementing the change to the program or the opportunity associated with implementing the change.
13. **Change owner:** Identify a member of the program management team to be responsible for ensuring that the change request is reviewed, that its impact on the program is analyzed, that it is submitted to the appropriate person for resolution, that once a decision is made on the change the decision is communicated to the requestor, and, if the change is approved, that this team member continues to track the change until its implementation is complete.

14. **Proposed resolution:** State the proposed resolution of the change after the change impact analysis has been conducted, such as approved, deferred, or rejected.

15. **Decision:** State whether the proposed resolution was finally approved, deferred, or rejected.

16. **Decision maker and actual resolution:** State who made the decision concerning the change request. Examples include:
    a. Change owner
    b. Program manager
    c. Change control board
    d. Governance Board

    Note if the change is approved, whether the proposed implementation of it is to be followed or whether a different implementation process is to be used. If the decision was to defer the change, continue to keep it on the log until it is either approved or rejected.

17. **Date:** State the date the change request was approved, deferred, or rejected.

18. **Subsequent impacts:** State any subsequent impacts on other controlled or configuration items (see the change management plan) as a result of implementation of this change request.

19. **Date closed:** State the date the change was implemented.

20. **Notes:** Use this field for any additional notes about the change.

Phase-Gate Review Agenda

One of the reasons programs continue beyond their usefulness is the lack of a systematic decision-making process. Phase gates occur at the end of each of the five phases described in the *Examination Content Outline*. In the *Standard for Program Management*—Third Edition (2013) they are discussed in section 6.2.4.5 with the explicit purpose of enabling the Governance Board to approve or "gate" from one phase to the next. As with any meeting, it is important to have an agenda that ensures an orderly and thorough meeting.

Phase-Gate Review Agenda Instructions

At different points in the program life cycle, phase-gate reviews will be held as described in the program's governance management plan. They enable the Governance Board to assess the program's progress in meeting its planned benefits according to the benefits realization plan. These reviews also are used to approve initiation of program components and to authorize closing or transitioning components, according to Project Management Institute's *Standard for Program Management*—Third Edition (2013). These meetings are used for decisions to move from one gate in the program management life cycle to another and to assess progress against specific criteria. They serve as go/no-go decisions to move to the next phase in the life cycle and to confirm that generally accepted best practices have been followed.

An agenda for these reviews is as follows:

PARTICIPANTS (NAMES/ORGANIZATION)

Date: _____

Time: _____

Place: _____

Program overview: The program manager presents a brief overview of the program as to why the program was undertaken and where it stands in the program management life cycle. The purpose is to enable participants to ask questions to see that objectives of the program and its components are still in alignment with the organization's strategic objectives.

Deliverables completed: The program manager describes the status of any completed deliverables. This section describes the program management processes that are being followed so participants can ask questions to ensure that best practices are being used.

Benefits realized: The program manager describes the status of the program benefits that have been realized to date. The purpose is to enable participants to determine if the benefits are being realized as stated in the benefits realization plan and are ones that support the program's business case.

Exit criteria satisfaction: The program manager describes the exit criteria for the previous phase and how it has been satisfied.

Risks: The program manager presents an overview of the risks that have affected the program since the last meeting or are expected to affect the program during the next phase in the life cycle. The purpose is to enable participants to ask questions to determine if the level of tolerance for risks is acceptable to the members of the Governance Board and the organization.

Expected date of the next meeting: The program manager states when he or she expects the next meeting will be held for planning purposes.

Decisions: This section documents the decisions made during this review.

Signatures of Participants

Governance Board chairperson _____

Governance Board member 1 _____

Governance Board member 2 _____

Governance Board member N _____

Program Review Agenda

In addition to the phase gate reviews, it is also useful to hold program reviews. The *Standard for Program Management*—Third Edition (2013) refers to these meetings as "periodic health checks," and they are also in the governance plan. Clearly, these review meetings need to be conducted during longer phases on longer programs but are less formal and are held between phase-gate reviews. These reviews also need an agenda for an orderly and thorough meeting.

Program Review Agenda Instructions

At different points in the program life cycle, program reviews will be held as described in the program's governance management plan. These meetings are used in addition to phase-gate reviews to assess overall program performance against outcomes and expected benefits and to determine if any preventive or corrective actions are needed. Also, since the phases in the life cycle may be long for many programs, these reviews provide a way to assess progress on a more frequent basis and keep the members of the Governance Board involved in the program to evaluate performance and to see if best practices are being followed. The program manager uses these reviews to focus on areas in the program rather than specifics and to describe overall status, concerns, and issues. The program manager also can use these reviews to discuss any alternatives with the Governance Board to improve opportunities for overall program success. Since these reviews are not go/no-go decisions, they generally are viewed in a more positive way and are conducted in a more informal manner.

An agenda for these reviews is as follows:

PARTICIPANTS (NAMES/ORGANIZATION)

Date: _____

Time: _____

Place: _____

Program overview: The program manager presents a brief high-level overview as to why the program was undertaken and where it stands in the program management life cycle. He or she should remind participants when the last phase-gate review and program review was held. Participants can ask questions to see that objectives of the program and its components are still in alignment with the organization's strategic objectives and to see whether the business case requires revision.

Status of milestones and deliverables: The program manager describes the status of completed deliverables and those that are expected to be completed in the near future as well as the status of key milestones. Participants can ask questions about the status of upcoming milestones and deliverables to see whether the schedule management plan is being followed and whether the schedule requires updating. They also can address customer satisfaction concerns, if any, and can see if the scope management plan is being followed.

Benefits status: The program manager describes the status of the program benefits that have been realized to date and are coming in the near future. Participants can ask questions to determine if the benefits are being realized as stated in the benefits realization plan and are ones that support the program's business case. They also can assess whether the benefits are ones that can be sustained.

Resource status: The program manager describes resource use in the program and whether additional resources are required or if any dedicated resources can be released. The program manager describes whether there are any problems in obtaining resources as agreed upon to support the program. Participants can ask questions to see that the program resource plan is being followed and to determine whether changes are required to the program manager's charter.

Component status: The program manager states whether any components have been transitioned to the customer or to ongoing operations and whether any new components are being added to the program. Participants can ask questions to see that closeout has occurred according to plans and that benefits are being sustained or, if components are being added, whether the program documentation is being updated. They can see if the program roadmap remains valid.

Budget status: The program manager provides an overview of the budget and discusses key earned value performance metrics, if earned value management is being used. Participants can ask questions to see that the financial plan is being followed. The Governance Board is responsible for ensuring programs are funded as descried in the program plan, and the Governance Board may be responsible for obtaining funding if it is provided by external sources.

Change request status: The program manager states the number of change requests that have been received and analyzed and explains whether those that have been approved have been implemented. The purpose is not only to describe the number of change requests to date but to enable participants to see that the change request process is being followed and to identify any items of concern that may affect future program performance.

Issue status: The program manager provides an overview of the status of issues on the program issue log. Participants then can ensure that the program issues log is being used and can ask questions about outstanding issues affecting the program. Also, the issue resolution process can be discussed to assess its effectiveness.

Stakeholder communications: The program manager uses the communications log and discusses communications with stakeholders to show the frequency of communication and to discuss any issues. He or she also describes other requests for information from stakeholders and whether these requests have led to the need to issue any other formal reports. Participants can ask questions to see that the program communications management plan and the program stakeholder engagement plan are being followed and can address any outstanding issues regarding program stakeholder engagement.

Risks: The program manager presents an overview of the risks that have affected the program since the last meeting or are expected to affect the program in the near future using the risk register. Participants can ask questions to determine if the level of tolerance for risks is acceptable and to see that the risk management plan, and the proposed risk responses are being followed. They can check on possible opportunities that have been identified and how they are being maximized.

Contract status: The program manager provides an overview of the contractor/supplier status. He or she describes completed deliverables, any issues, whether agreements and contract terms and conditions are being followed, and the need for additional contracts in other areas. Participants can ask questions to see that the procurement management plan and contracts management plan are being followed.

Lessons learned: The program manager discusses lessons learned in the program to date using the lessons learned log. Participants can ask questions to see if program management processes and procedures require change, are being used as intended, and are adding value to the program. They also may see opportunities to leverage these lessons learned to other programs and projects under way in the organization or to transfer knowledge assets from other areas to this program. The emphasis during this part of the review is to review these lessons learned to improve overall program success.

Additional comments: The program manager provides additional comments about the program, and participants ask questions in other areas as appropriate.

Expected date of the next meeting: The Governance Board members state when they expect to hold the next meeting for planning purposes.

Decisions: This section documents any decisions made during this review.

Signatures of Participants

Governance Board chairperson   _____

Governance Board member 1   _____

Governance Board member 2   _____

Governance Board member 3   _____

Stakeholder 1   _____

Stakeholder 2   _____

Stakeholder 3   _____

Issue Escalation Process

Questions surrounding unplanned occurrences must be referred to the program from its components (projects and other work), and from the program to organizational executive management. Seasoned program managers know when it is and when it is not important to make those referrals and how to do them properly. This is the purview of the issue escalation process noted in the governance plan and in section 6.2.4.9 in the *Standard for Program Management*—Third Edition (2013). The issue escalation process is used as the program is executed and monitored and controlled.

Issue Escalation Process Instructions

Purpose: Each program will have issues, or unplanned events, actions, or disputes that may impact program areas and that will need resolution, following the program's governance plan and using the program issues register or log. One of the program manager's responsibilities is to resolve these issues that are escalated from program components through a synergistic approach to enable the program's benefits to be realized. It is also a key part of the program charter and program management plan to describe the identified risks and issues at the beginning of the program so they do not turn into problems later during execution. The program manager proactively strives to anticipate issues and manage them effectively.

An issue resolution process can facilitate the steps to follow when there are issues that cannot be resolved by a component manager, a team member, or the program manager. It is important to recognize that issues may affect any aspect of the program or other programs.

In the *Standard for Program Management*—Third Edition (2013), this issue escalation process part of the Governance domain in 6.2.4.9.

This document describes an issue escalation process to follow:

1. Set up the program issues log or register at the beginning of the program to identify the issue and to track it throughout the process. Assign a number to the issue and link it to the program work breakdown structure (PWBS) number. Any issues noted in the business case, program mandate, and the program charter should be added to the program issues register once the program is officially initiated.
2. Determine if a change request is required for the issue, and if so, have someone on the program management team prepare it. Ideally, the change request should be prepared by the person who first identified the issue.
3. Determine the issue's priority and identify someone on the program management team who will be responsible for tracking the progress of the issue through its resolution and subsequent implementation. Often, the program management office handles the issue log for the program.
4. Determine whether a feasibility study may be required to help analyze the issue and determine the most appropriate way to resolve it, and if so, identify a program management team member to conduct this study and prepare a plan for it so it can be completed in a timely manner. Use this feasibility study as part of the overall issue analysis process to determine if additional funding is required for the issue resolution process or to implement corrective action.
5. Inform the Governance Board that an issue has been identified through a regularly scheduled status report or by other means if it is an issue that has been identified to have a major impact on the program or the

organization, in case it cannot be resolved by the program manager. The Governance Board should be informed as soon as possible in case a special meeting may need to be held to resolve the issue.

6. Review the program manager's charter to see if he or she has the authority to resolve the issue. Typical resolution methods are to accept the issue and not require other changes or to change the program management plan and other associated documents because of the issue. The issue escalation process and the people to be involved typically are covered in the program communications management plan and in the Governance plan.

7. Determine the stakeholders that will be affected by the issue and engage them as appropriate in the issue resolution process. Identify possible stakeholders who may be affected by reviewing the type of issue as noted in the program issues log. Update the stakeholder register accordingly to ensure all stakeholder concerns have been addressed.

8. If the program manager cannot resolve the issue effectively, escalate it to the Governance Board for resolution. Even if the program manager has the authority to resolve the issue, it may be one that affects other programs or projects or the entire organization, so the program manager may decide to involve the Governance Board because of the issue's scope and impact.

9. If the issue is escalated to the Governance Board, the program manager should describe the importance of the issue. It may be appropriate to have a member of the program management team, or someone from the program management office, to facilitate the meeting so a decision concerning the issue can be made quickly and to see that the Governance Board members' questions are answered during the meeting. The program manager then should communicate the resolution of the issue after the meeting to affected stakeholders, and the team member responsible for the program issues log should note the decision in the log.

10. Ensure that once a resolution for the issue has been determined that corrective actions are taken. Also determine whether preventive actions are required so the issue does not surface in the future.

11. Document the issue and its resolution as part of the lessons learned register.

12. In the next meeting with the Governance Board, state that the issue has been resolved and the corrective actions that have been taken. Note that it is part of the lessons learned register in case there are other projects or programs under way that could benefit from learning about this issue and its resolution.

13. In a status report that is distributed to stakeholders and members of the Governance Board, state that the issue was resolved and describe its resolution.

14. Ensure that as part of the transition plan after the program is officially closed, there is customer support in case a comparable issue arises later after the program deliverables are completed.

Component Initiation Request

Programs consist of projects as well as management effort and infrastructure—known generically as components and non-project work. Can just anyone at any time declare a new project to begin using up limited resources? Let's hope not! Business partners can even disagree on whether they agreed to start a project. Decisions regarding new components are best handled through a component initiation form so there is no uncertainty. This form is prepared in advance because the Governance Board must approve the initiation requests as noted in section 6.2.12 in the *Standard for Program Management*—Third Edition (2013).

Component Initiation Request Instructions

The component initiation request includes the following:

Purpose: A brief introductory statement defining the purpose of the component initiation request, such as:

> The component initiation request describes the need, feasibility, and justification for the component to be part of the program.

> Components include both projects and non-project or operations work that is part of the program. In most cases, the program manager prepares the component initiation request with input from other key stakeholders and the program sponsor. It may be provided by the client or the funding organization. Approval by the Governance Board typically is required especially if new governance structures are required or if organizational resources are required. This approval serves as the mandate for the component to be part of the program.

> While some components already are under way when the program is initiated, other components will become part of the program at different phases in the life cycle, except during the closing phase. They are generally included in the program roadmap and the program management plan, with explicit criteria to show when components are to be initiated. As the work on the program ensues, other components that were not previously identified may be needed. The component initiation request is a document to secure formal approval for program components. Components to be part of the program must meet the organization's approved selection criteria. These criteria typically are stated in the program's governance plan. Once approved, a charter for the component is prepared and approved, a project manager or operations manager is assigned, and the work begins. The actual work to be done then follows the guidelines in PMI's PMBOK® Guide, unless this component is a sub-program. Then it would follow guidelines in the *Standard for Program Management*—Third Edition (2013).

In the Project Management Institute's *Standard for Program Management—Third Edition* (2013), component initiation requests are described in the governance plan in section 6.2.12.

Component business case: Each component requires a business case to show how its expected benefits will be more significant if it is part of the program rather than managed as a stand-alone project or operational activity. This section describes the business case for the component. The component's expected results should complement those of the program and should be aligned with the organization's strategic plan and goals. The component also may be inter-related in various ways to other components in the program. Its financial indices must be detailed along with internal and external interfaces. It may require compliance with the program's quality plan. If feasibility studies or alternative analyses were conducted, they should be noted in this section.

Project manager or operations manager: This section identifies the project manager or operations manager for the component and provides contact information.

Component sponsor: This section identifies the sponsor for the component and provides his or her contact information. In most cases the sponsor is the program manager.

Stakeholder communications: Once a component is approved to be part of the program, stakeholders will need to be notified. This section states those stakeholders who will have an active role with the component, will influence the outcome of the component, or will be interested in certain aspects of the component, and describes how these stakeholders will be notified of component approval.

Component governance structure: While the program has a governance management plan and uses a Governance Board, components also require a governance structure at the program level. This section states the processes and procedures to be used to monitor and track component progress. It describes gate reviews and performance reviews to be held and specifies status reports to be prepared at the component level. It also notes that audits may be conducted at scheduled and nonscheduled times.

Resource requirements: This section provides a high-level estimate of the resources required for the component. It may be necessary to transfer resources from existing components already under way in the program to the proposed component, especially if it has a higher priority than that of the existing work. This section states why the various resources are needed and when resources need to be available.

Change requests: Once a component is approved, change requests will be required. This section states the various change requests that will be needed for the specific component when it is approved.

Program documentation changes: Each time a new component is approved, program documents will require updates as noted by change requests. This section describes each of the program documents that will be updated and states the program team member who is responsible and when the updates are scheduled to be completed.

Approvals: This section contains the written approval of the component initiation request by the program manager, program sponsor, program management office, members of the program's Governance Board, or any other key stakeholders as appropriate.

Component Initiation Request Template

<Insert Program Name>
Component Initiation Request

Program name:	
Program manager:	PM's email address here as a hyperlink
Program sponsor:	
Proposed component start date:	
Proposed component end date:	
Date prepared:	
Program number:	
Revision history:	
Business unit:	

A. PURPOSE

A brief introductory statement defining the purpose of the component initiation request, such as:

> The component initiation request describes the need, feasibility, and justification for the component to be part of the program.

_____

_____

_____

B. COMPONENT BUSINESS CASE

This section describes the business case for the component and describes why it is more effective to include this component in the program rather than managing it as a stand-alone project or operational activity.

_____

_____

_____

C. PROJECT MANAGER OR OPERATIONS MANAGER

This section states the name of the project manager or operations manager and provides his or her contact information.

_____

_____

_____

D. COMPONENT SPONSOR

This section identifies the sponsor for the component and provides contact information.

_____

_____

_____

E. STAKEHOLDER COMMUNICATIONS

This section states those stakeholders who will be notified when the component initiation request is approved.

_____

_____

_____

F. COMPONENT GOVERNANCE STRUCTURE

This section describes the governance structure for the program's components.

_____

_____

_____

G. RESOURCE REQUIREMENTS

This section provides a high-level estimate of the resources required for the component and describes why they are needed and when they need to be available.

_____

_____

_____

H. CHANGE REQUESTS

This section states the various change requests that will be needed when the component is approved.

_____

_____

_____

I. PROGRAM DOCUMENTATION CHANGES

This section describes each of the program documents that will be updated when the component is approved, the program team member who is responsible for the update, and when it is expected to be completed.

_____

_____

_____

J. APPROVALS

This section contains the approval of the component initiation request by the program manager, program sponsor, program management office, members of the program's Governance Board, and other key stakeholders as required.

SIGNATURES AND DATE APPROVAL OBTAINED

Program manager   _____

Program sponsor   _____

Governance Board member 1   _____

Governance Board member 2   _____

Governance Board member N   _____

Stakeholder 1   _____

Stakeholder 2   _____

Stakeholder N   _____

Component Transition Request

As noted, various components will be initiated during a program. These components will transition to ongoing operations, such as to a product or customer support group, to customers, or to users at different times during the course of the program. The Governance Board will approve the requests to transition these components. In the *Standard for Program Management*—Third Edition (2013), a decision from the Governance Board to transition the component is required as stated in section in section 6.2.12.

Component Transition Request Instructions

The component transition request includes the following:

Purpose: A brief introductory statement defining the purpose of the component transition request, such as:

> The component transition request shows that the component's benefits have been realized, its business case has been sufficiently satisfied, and the component should be transitioned to ongoing operations.

Components include both projects and non-project or operations work that is part of the program. Different components will finish ahead of others as noted by completion of project deliverables, completion of program-level milestones, and achievement of benefits.

In most cases, the project manager or operations manager prepares the component transition request with input from other key stakeholders and the program manager. It may be provided by the client or the funding organization. The Governance Board typically approves this request. It serves as the mandate to ensure the knowledge assets, responsibilities, and benefits are ready to be handed over to ongoing operations.

It is noted that components also will transition from one phase in their life cycle to another. This form, though, is used for formal approval for component closure, rather than for approval to transition to the next phase in the life cycle, as that is covered as part of the program's governance procedures. This transition is the last phase in the component's life cycle.

This form is not used to terminate a component when it is apparent that the component does not support the program's objectives and does not contribute to the program's benefits; termination is handled by a termination request.

In the Project Management Institute's *Standard for Program Management—Third Edition* (2013), the component transition request is described in section 6.2.12, and in reviewing the recommendation for transition, the Governance Board evaluates the component's business case, ensures stakeholders involved with this component have received communications relative to its transition, ensures compliance with the program's quality plan if applicable, assesses organization or program lessons learned, and confirms practices for official closure have been followed.

Reasons for request: This section describes the reasons for the transition request. For example, the component has achieved its specific benefits, has met its requirements as stated in the business case, has completed all of its deliverables, and/or has met all program-level milestones.

Resources released: Once the component transition request is approved, resources can be reallocated to other parts of the program or to the performing organization. This section states the available resources for planning purposes by the program manager, program sponsor, and portfolio manager, as appropriate.

Closure requirements: In order for the component to transition to ongoing operations, the program's closing policies and procedures must be met. This section states that all closure activities are complete, such as archiving records, returning customer property, and documenting lessons learned, as defined by the program management plan and governance plan.

Stakeholder communications: Each component will have its own stakeholders, in addition to stakeholders at the program level. This section describes how the affected or interested stakeholders will learn that the component is formally closed.

Change requests: Once a component is approved for transition, change requests will be required. This section states the various change requests that will be needed as a result of approval of the transition request.

Program documentation changes: This section states the requirements to update program-level documentation because of a component transition request, such as the program management plan, program resource plan, decision log, and program roadmap. It describes the program management team member who is responsible for updating the artifacts, and the planned date the updates will be completed.

Project completion closure certificate: This section describes who will prepare the certificate noting that project completion has occurred and who must sign off on it once the transition request is approved.

Approvals: This section contains the written approval of the component transition request by the program manager, program sponsor, members of the program's Governance Board, or any other key stakeholders as appropriate.

Component Transition Request Template

<Insert Program Name>
Component Transition Request

Program name:	
Program manager:	PM's email address here as a hyperlink
Project or operations manager:	
Program sponsor:	
Component start date:	
Proposed component end date:	

Program number:	
Date prepared:	
Prepared by:	

A. PURPOSE

A brief introductory statement defining the purpose of the component transition request, such as:

> The component transition request shows that the component's benefits have been realized, its business case has been sufficiently satisfied, and the component should be transitioned to ongoing operations, customers, or users.

_____

_____

_____

B. REASONS FOR REQUEST

This section describes the reasons for the transition request.

_____

_____

_____

C. RESOURCES RELEASED

This section states the resources that will be released once the transition request is approved in order that they may be used on other program components or elsewhere in the organization.

_____

_____

_____

D. CLOSURE REQUIREMENTS

This section describes the program's closure requirements for its components to ensure that all of them have been completed.

_____

_____

_____

E. STAKEHOLDER COMMUNICATIONS

This section states those stakeholders who will be notified when the component transition request is approved.

F. CHANGE REQUESTS

This section describes the change requests that will be needed once the transition request is approved.

G. PROGRAM DOCUMENTATION CHANGES

This section describes each of the program documents that will be updated when the component transition request is approved, the program management team member who is responsible for the update, and when it is expected to be completed.

H. PROGRAM COMPLETION CLOSURE CERTIFICATE

This section states who will prepare the certificate noting the completion has occurred and who must sign off on it.

I. APPROVALS

This section contains the approval of the component transition request by the program manager, program sponsor, members of the program's Governance Board, and other key stakeholders as required.

SIGNATURES AND DATE APPROVAL OBTAINED

Program manager

Program sponsor

Governance Board member 1

Governance Board member 2

Governance Board member N

Stakeholder 1

Stakeholder 2

Stakeholder N

Program Closure Recommendation

Program closure is clearly a major governance issue. Certain issues must be settled in the process, such as, for example, whether continuing benefits are transitioned to ongoing operations or become a part of another program. In the *Standard for Program Management*—Third Edition (2013), this recommendation is discussed in section 6.2.13, Program Governance, in the Governance Domain.

Program Closure Recommendation Instructions

The program closure recommendation includes the following:

Purpose: A brief introductory statement defining the purpose of the program closure recommendation, such as:

> The program closure recommendation states that all program benefits have been realized, and all deliverables have been completed, so the program can be officially closed.

The program manager prepares this recommendation and presents it to the Governance Board as the program then is in the program closure phase in its life cycle. Before the recommendation is prepared, the program manager should meet with each of the key stakeholders to ensure they support this recommendation. The Governance Board documents its recommendation to close the program. This decision should be included in the governance decision register. This decision then is given to the program sponsor, who officially closes the program.

At this point, the program manager states that all components have been completed and have been closed according to the program's closure policies, all lessons learned have been documented, all documents have been archived, all resources have been reassigned, and any property provided by others has been returned.

The program manager provides metrics as part of the recommendation to show that expected program benefits have been realized, deliverables have been completed, and the program's goals have been met.

Any work that may need to be done to transition the program to ongoing operations is described as part of the recommendation.

Reasons for request: This section describes the reasons for the program closure recommendation. For example, it describes that the program's benefits have been realized, deliverables have been completed, and objectives have been met. The program manager may note any forecasts that have been prepared concerning the ongoing value of the benefits from the program. This request also is prepared if the program must be closed prematurely and states why it is to be closed.

Resources released: Once the program closure recommendation is approved, any remaining resources can be reallocated to other parts of the performing organization. This section states the available resources for planning purposes by the program sponsor and portfolio manager, as appropriate.

Disposition of property: This section states that any customer or performing organization property provided to the program has been returned.

Recommended changes: At the time the program officially is closed, during the lessons learned review session, the program management team may have noted some changes that could be beneficial to other programs and projects under way in the organization. This section describes any of these recommendations for consideration.

Continued benefits realization: This section describes requirements for any continued benefits realization once the program is closed, and the benefits are transitioned to the customer, users, a product or customer support group in the organization, or another program under way in the organization. This section describes any ongoing activities that may be required to ensure the benefits continue to be sustained and recommends any tracking or monitoring of these benefits as required.

Stakeholder communications: This section describes how the affected or interested stakeholders will learn that the program is formally closed.

Program completion closure certificate: This section describes who will prepare the certificate noting that program completion has occurred and who must sign off on it once the closure recommendation is approved.

Approvals: This section contains the written approval of the program closure recommendation by the program manager, program sponsor, members of the program's Governance Board, and any other key stakeholders as appropriate.

Program Closure Recommendation Template

<Insert Program Name>
Program Closure Recommendation

Program name:	
Program manager:	PM's email address here as a hyperlink
Program sponsor:	
Actual start date:	
Proposed closure date:	
Program number:	
Revision history:	
Business unit:	

A. PURPOSE

A brief introductory statement defining the purpose of the program closure recommendation, such as:

> The program closure recommendation states that all program benefits have been realized, and all deliverables have been completed, so the program can be officially closed.

_____

_____

_____

B. REASONS FOR REQUEST

This section describes the reasons for the program closure recommendation.

_____

_____

_____

C. RESOURCES RELEASED

This section states the resources that will be released once the closure recommendation is approved in order that they may be used elsewhere in the organization.

_____

_____

D. DISPOSITION OF PROPERTY

This section states that any customer or performing organization property provided to the program has been returned.

_____

_____

_____

E. RECOMMENDED CHANGES

This section describes any recommended changes in program management practices based on lessons learned in this program that could be beneficial for other programs and projects in the organization.

_____

_____

_____

F. CONTINUED BENEFITS REALIZATION

This section describes requirements for any continued benefits realization once the program is closed, and the benefits are transitioned to a customer, users, a product or customer support group in the organization, or another program.

_____

_____

_____

G. STAKEHOLDER COMMUNICATIONS

This section states those stakeholders who will be notified when the program closure recommendation is approved.

_____

_____

_____

H. PROGRAM COMPLETION CLOSURE CERTIFICATE

This section describes who will prepare the certificate noting that program completion has occurred and who must sign off on it.

_____

_____

_____

I. APPROVALS

This section contains the approval of the program closure recommendation by the program manager, program sponsor, members of the program's Governance Board, and other key stakeholders as required.

SIGNATURES AND DATE APPROVAL OBTAINED

Program manager _____

Program sponsor _____

Governance Board member 1 _____

Governance Board member 2 _____

Governance Board member N _____

Stakeholder 1 _____

Stakeholder 2 _____

Stakeholder N _____

Audit Plan

Audits can be conducted by elements internal to the program or external to it. In contrast to meetings where progress and status are reported, audits focus more on the supporting data and record keeping. The objectives and timing of audits are in the audit plan described in the Governance Domain in section 6.6.4 in the *Standard for Program Management*—Third Edition (2013). The Standard notes the Governance Board may assume responsibility for ensuring that the program manager and his or her team are prepared for audits. To help best prepare for these audits, a plan is helpful.

Audit Plan Instructions

The audit plan includes the following:

Purpose: A brief introductory statement defining the purpose of the audit plan, such as:

> The audit plan describes the process that will be followed to periodically review the overall performance of the program at all levels to enhance effectiveness and ensure that the program benefits are being delivered as set forth in the benefits realization plan and that the program is being executed as stated in the program management plan.

Audits are common in programs. They are necessary for a number of reasons, such as on finances, quality, management processes and practices, and they should be viewed in a positive manner rather than as necessary for compliance or for pinpointing specific problems and associating them to specific individuals.

Audits may be internal or external. Ideally, the program manager should request internal audits at various times in the program and should establish a process so that everyone on the team views the auditors in a positive way, with the audit results then used to improve the program's overall effectiveness.

This plan is an iterative document and should be reviewed periodically by the program management team and other key stakeholders, especially the members of the Governance Board, as the work of the program continues throughout the various phases in its life cycle. It also should be reviewed after each audit is conducted.

In the Project Management Institute's *Standard for Program Management—* Third Edition (2013), the audit plan is noted in section 6.6.4. The audit plan is a subsidiary plan to the overall program management plan.

Goals: This section states the goals for program audits. Examples are to examine the financial status of the program; to review the effectiveness of the program management processes, procedures, and guidelines that are being used; to determine whether or not there is compliance with specific standards and regulations; and to assess whether there may be evidence of fraud or mismanagement. The specifics of each program dictate the various types of audits that may be required; however, each program on a periodic basis should conduct audits to assess overall program management effectiveness and to determine whether changes in procedures, processes, and guidelines are needed to help ensure overall program success.

Roles and responsibilities: This section describes the roles and responsibilities of the program management team and other key stakeholders in preparing for an audit and during an audit. While the auditors, whether internal or external, will prepare a plan for the audit, the program manager and other members of the program management team will also need to plan for the audit and must participate in it. A number of people on the team and other stakeholders may need to be interviewed, and the program manager will need to ensure that the auditors have access to program documentation. The program manager should set the stage for the audit with the team to promote an atmosphere of cooperation. Everyone on the team should view the results of the audit as ones that can improve overall program effectiveness. The program manager must serve as the facilitator for the auditors.

Schedule: This section presents the schedule for program audits. Although some audits will occur on a random basis and cannot be scheduled, others can be part of the program master schedule. The program manager ideally should request audits at certain times for a self assessment during the program's life cycle, such as at key milestones in the program, at the end of a phase-gate review, or before a scheduled program review to help prepare for such reviews and to have other suggestions to note to the Board during these review meetings.

Audit results: The auditors, whether internal or external, will present a report that documents the results and provides recommendations. This section describes how these audit results will be documented and made available to key stakeholders. Ideally, a practice of "no secrets" will be followed. This means that the results should be available in an easily accessible format so stakeholders, especially those on the program management team, can review the suggested recommendations. This section also describes how the program management team will implement the suggested recommendations. Some recommendations may be mandatory, especially if the audit is conducted to determine compliance with regulations or standards. If the program management team believes it cannot implement a mandatory recommendation, this section describes the process to request a waiver. It also states the process to follow if the program management team wants to implement a recommendation at a different time than specified in the audit report. The program manager may need to escalate these requests to the Governance Board for approval. Additionally, this section describes the process to update other plans and documents as needed based on the findings and recommendations in the audit report. Each recommendation will require a change request and should be handled through the program's change control system.

Monitoring and tracking: This section describes the approach the program management team will use to monitor and track the implementation of the recommendations from each audit. It describes the use of an information repository and/or a log or register to assist in this process. In subsequent audits, this log then can be provided to the auditors.

Approvals: This section contains the written approval of the audit plan by the program sponsor, program manager, program management office, members of the Governance Board, and any other key stakeholders as appropriate.

Audit Plan Template

<Insert Program Name>
Audit Plan

Program name:	
Program manager:	PM's email address here as a hyperlink
Program sponsor:	
Actual start date:	
Approved end date:	
Program number:	
Revision history:	
Business unit:	

A. PURPOSE

A brief introductory statement defining the purpose of the audit plan, such as:

The audit plan describes the process that will be followed to periodically review the overall performance of the program at all levels to enhance effectiveness, ensure that the program benefits are being delivered as set forth in the benefits realization plan, and that the program is being executed as stated in the program management plan.

_____

_____

_____

B. GOALS

This section states the goals for the audits.

_____

_____

_____

C. ROLES AND RESPONSIBILITIES

This section describes the roles and responsibilities of the program management team and other key stakeholders in preparing for an audit and during an audit.

_____

_____

_____

D. SCHEDULE

This section presents a schedule for program audits, even though some will occur on a random basis.

_____

_____

_____

E. AUDIT RESULTS

This section describes how the audit results will be documented and made available to program stakeholders. It describes a process that the program management team will follow if the team believes a mandatory recommendation from an audit cannot be implemented or can be implemented but at a later time than that recommended. It describes the process to update plans and other documents as needed based on the audit findings and recommendations.

_____

_____

_____

F. MONITORING AND TRACKING

This section describes the approach the program management team will use to monitor and track the implementation of the recommendations from each audit.

_____

_____

_____

G. APPROVALS

This section contains the approval of the audit plan by the program sponsor, program manager, program management office, members of the Governance Board, and other key stakeholders as required.

SIGNATURES AND DATE APPROVAL OBTAINED

Program manager _____

Program sponsor _____

Program management office director _____

Governance Board chairperson _____

Governance Board member 1 _____

Governance Board member 2 _____

Governance Board member N _____

Stakeholder 1 _____

Stakeholder 2 _____

Stakeholder N _____

Audit Report

Audits provide important and more in-depth verification of actual program execution and performance. This report format can ensure completeness and consistency of the reports. It supports the audit plan noted in section 6.6.4 in the *Standard for Program Management*—Third Edition (2013).

Audit Report Instructions

The Audit Report includes the following:

Purpose: A brief introductory statement defining the purpose of the Audit Report such as:

> The Audit Report presents an objective assessment of the performance of the program. The report's findings and recommendations serve to enhance program effectiveness and ensure that the program benefits are being delivered as set forth in the benefit realization plan and that the program is being executed as stated in the program management plan.

> Audits are common on programs. They are necessary for a number of reasons and should be viewed in a positive manner rather than considering them as necessary for compliance or for pinpointing specific problems and associating them to specific individuals. The Audit Report is an objective report of findings and provides recommended preventive or corrective actions to follow. It also is used to ensure program management processes are being followed and can be a way for checks and balances based on other views for governance decisions.

> According to the In the Project Management Institute's *Standard for Program Management*—Third Edition (2008), the audit plan is described in section 6.6.4 as part of the Governance domain. The Audit Plan sets the stage for the report.

Background Information:

Time Period of the Audit: This section states the time the audit began and when the final report was prepared.

Place: This section states where the audit was conducted.

Audit Participants: This section states the names of the people on the audit team and provides their contact information for possible questions.

Type of Audit: This section states the type of audit that was conducted: internal or external. Note whether the audit was conducted according to a schedule in the program's master schedule based on the Audit Plan or whether it was random.

Reason for the Audit: This section describes why the audit was conducted such as to concentrate on a component that may be in trouble or because the overall program may be in trouble, to look at specific processes and procedures to see whether or not they are being followed, for compliance purposes with specific regulations or procedures, to examine the overall financial status of the program, to see if benefits are being realized as stated in the benefits realization plan, or for other reasons.

Program Areas Affected: This section describes the various areas of the program that were involved in the audit: e.g., the entire program or only certain components of it.

Methodology: This section describes how the audit was conducted. For example, it notes whether documents were reviewed, interviews were held, surveys were used, focus groups were used, and/or other methods.

Findings: This section presents the findings from the audit. The findings should be presented in an objective way as they are the basis for the recommendations from the audit. They are the synthesis of all of the data that were reviewed. Findings should be carefully worded to reflect the audit team's observation, and they should be phrased as constructively presented problem statements. Appropriate background information that is needed to understand the findings should be included. Causes are observations that support the findings, and their identification is helpful in making constructive recommendations. Consequences list the problem results of a finding. Findings should represent issues for the entire program or for the components being audited and should have the broadest possible application.

Here is an example:

Finding (problem statement): Each status report that is prepared involves development of completely new material.

Probable cause: Reusable materials are not available.

Business consequence: Reports that are prepared may not be completed on schedule or may not contain the information of interest to the program's stakeholders.

Recommendation: Prepare a Communications Management Plan that describes the types of reports to prepare and provide them to various program stakeholders, and the data each report should include. Ask the Program Management Office to take the lead in providing a sample report template to use throughout the program.

In preparing findings, keep the specific audit goals and appraisal scope in mind. Avoid the following:

■ Moot issues
■ Findings based on hearsay alone
■ Broad generalizations

Recommendations: This section states the recommendations based on the audit findings. Use categories if possible to group the recommendations and present them in priority order. If the audit is for compliance purposes, note if some of the recommendations are mandatory.

Appendices: Include appendices such as the following:

Persons Interviewed: List the people who were interviewed during the audit and the date of each interview

Documents Reviewed: List the documents that were reviewed during the audit

Approvals: This section contains the written approval of the Audit Report by the lead Auditor.

Audit Report Template

<Insert Program Name>
Audit Report

Program name:	
Program manager:	PM's email address here as a hyperlink
Program sponsor:	
Actual start date:	
Approved end date:	
Program number:	
Revision history:	
Business unit:	

A. PURPOSE

A brief introductory statement defining the purpose of the Audit Report such as:

> The Audit Report presents an objective assessment of the performance of the program. The report's findings and recommendations serve to enhance program effectiveness, ensure that the program benefits are being delivered as set forth in the benefit realization plan, and that the program is being executed as stated in the program management plan.

_____

_____

_____

B. BACKGROUND INFORMATION

This section provides background information about the audit. It should describe when the audit began and was completed, where the audit was conducted, who participated on the audit team, the type of audit that was conducted, the reason the audit was conducted, and the program areas involved in the audit.

_____

_____

_____

C. METHODOLOGY

This section describes how the audit was conducted.

_____

_____

_____

D. FINDINGS

This section presents the findings from the audit.

_____

_____

_____

E. RECOMMENDATIONS

This section states the recommendations based on the audit findings, in priority order, with notes if there are any mandatory recommendations if the audit was conducted for compliance purposes.

_____

_____

_____

F. APPENDICES

This section lists appendices such as the name of people who were interviewed, and the documents that were reviewed.

_____

_____

_____

G. APPROVALS

This section contains the approval of the Audit Report by the lead Auditor.

SIGNATURES AND DATE APPROVAL OBTAINED

Lead Auditor   _____

Quality Management Plan

Program quality planning and standards are discussed in the Governance Domain in section 6.2.9 in the *Standard for Program Management*—Third Edition (2013) since quality planning is essential to the program's components and any sub-programs, noting also the importance of quality planning at the program level. A quality management plan is a best practice to establish minimum standards for quality planning, quality control, and quality assurance for the program's components.

Quality Management Plan Instructions

The quality management plan will include the following:

Purpose: A brief introductory statement defining the purpose of the quality management plan, such as:

> The quality management plan establishes mechanisms for program quality and describes the program's quality requirements that cross the various components of the program. It also provides requirements to assist the program components as project managers in the program prepare their quality management plans.

> It should be noted that the program's quality management plan does not replace the quality management plans to be prepared by each of the project managers working in the program. Instead, its purpose is to state specific quality standards and policies, as well as oversights, designed to ensure that the program realizes the benefits it is to achieve. This plan also serves to ensure that there is consistent application of specific quality requirements across the various components (projects and other work) in the program.

Organizational quality standards: This section states the performing organization's quality standards and policies. It also should note quality standards and policies of the program's customers. If the program is a consortium, partnership, or joint venture, the program management team will need to set specific standards and policies to be followed in the program. Note that standards are established norms, methods, policies, and practices. The quality policy sets forth the intended direction of the organization with regard to quality.

Overview of the program's benefits: This section presents an overview of the benefits the program is designed to achieve. It can show a link to the program's benefits realization plan. With a complete understanding of the scope and nature of the benefits from the program, the program manager then can more effectively design appropriate quality standards and measures.

Program quality planning tools and techniques: This section describes the tools and techniques the program management team will use as they prepare the program's quality management plan. Examples of tools and techniques that may be used include cost-benefit analysis, benchmarking, checklists, and determination of the total cost of program quality.

Program quality assurance and quality control tools and techniques: This section states the quality assurance and quality control techniques the program team will use for program-level quality activities.

Requirements for quality planning, quality assurance, and quality control for the program's components: This section describes the minimal requirements each of the program's components must follow concerning quality planning, quality assurance, and quality control. It specifies the minimal quality standards and testing standards for the components. The purpose is to ensure that the quality management activities of the program components support the overall quality policy and standards at the program level.

Program-level roles and responsibilities: This section describes the program manager's roles and responsibilities for maintaining quality standards as well as for quality assurance and quality control. Quality assurance responsibilities may include audits of the program's overall commitment to quality management or an audit of a specific component's quality management activities to ensure appropriate standards are used, and there is compliance as required with regulations. These responsibilities also focus on continuous process improvement for the program. Quality control responsibilities may include methods to monitor and record results of executing activities to assess performance and to recommend preventive or corrective actions. Note that both quality assurance and quality control are performed throughout the program.

Quality metrics: This section describes quality metrics that will be used in quality assurance and quality control. These metrics assist in comparing actual practices to planned practices to ensure program requirements are met successfully, and rework is not required. These metrics show compliance as well with the program's standards and any applicable regulations.

Approvals: This section contains the written approval of the quality management plan by the program sponsor, program manager, program management office, members of the Governance Board, and other stakeholders.

Quality Management Plan Template

<Insert Program Name>
Quality Management Plan

Program name:	
Program manager:	PM's email address here as a hyperlink
Program sponsor:	
Actual start date:	
Approved end date:	
Program number:	
Revision history:	
Business unit:	

A. PURPOSE

A brief introductory statement defining the purpose of the quality management plan, such as:

> The quality management plan establishes mechanisms for program quality and describes the program's quality requirements that cross the various components of the program. It also provides requirements to assist the program components as project managers in the program prepare their quality management plans.

_____

_____

_____

B. ORGANIZATIONAL QUALITY STANDARDS

This section states the performing organization's quality standards and policies. It also should note the customer's quality standards and policies. If the program is a joint venture, the program management team will need to set these standards and policies.

_____

_____

_____

C. OVERVIEW OF THE PROGRAM'S BENEFITS

This section presents an overview of the benefits the program is to achieve in order to more effectively design program quality standards and metrics. A link to the program benefits realization plan may be included.

_____

_____

_____

D. PROGRAM QUALITY PLANNING TOOLS AND TECHNIQUES

This section describes the tools and techniques the program management team will use as they prepare the program's quality management plan.

_____

_____

_____

E. PROGRAM QUALITY ASSURANCE AND QUALITY CONTROL TOOLS AND TECHNIQUES:

This section states the quality assurance and quality control techniques the program management team will use for program-level quality activities.

_____

_____

_____

F. REQUIREMENTS FOR QUALITY PLANNING, QUALITY ASSURANCE, AND QUALITY CONTROL FOR THE PROGRAM'S COMPONENTS

This section describes the minimal requirements each of the program's components must follow concerning quality planning, quality assurance and quality control in order that the quality management activities of the program components support the overall quality policy and standards at the program level.

_____

_____

_____

G. PROGRAM-LEVEL ROLES AND RESPONSIBILITIES

This section describes the program manager's roles and responsibilities for maintaining quality standards as well as for quality assurance and quality control at the program level. These roles and responsibilities are performed throughout the program.

_____

_____

_____

H. QUALITY METRICS

This section describes quality metrics that will be used in quality assurance and quality control at the program level to ensure program requirements are met successfully, rework is not required, and there is compliance with the program's standards and any applicable regulations.

_____

_____

_____

I. APPROVALS

This section contains the approval of the quality management plan by the program sponsor, program manager, program management office, members of the Governance Board, and other key stakeholders.

SIGNATURES AND DATE APPROVAL OBTAINED

Program manager	_____
Program sponsor	_____
Program management office director	_____
Governance Board chairperson	_____
Governance Board member 1	_____
Governance Board member 2	_____
Governance Board member N	_____
Stakeholder 1	_____
Stakeholder 2	_____
Stakeholder N	_____

Knowledge Management Plan

One of the most important assets of any organization is its corporate knowledge. This asset does not happen or grow accidentally. A knowledge management (KM) plan is therefore a key to success.

Knowledge Management Plan Instructions

The knowledge management (KM) plan includes the following:

Purpose: A brief introductory statement defining the purpose of the knowledge management plan, such as:

> The knowledge management plan shows how knowledge management can improve the effectiveness and efficiency of overall program management, as it describes how people will be connected to one another in the program and to information generated throughout the course of the program.

> This plan should be prepared early in the program because through a commitment to KM by the program management team, greater efficiencies should result. People on the program team will be able to access information

prepared by others with whom they may not have direct contact or even know. People also will be able to locate subject matter expertise that could help with key program issues. The emphasis is on use of KM to avoid reinvention, promote reuse, use less experienced staff on issues and problems for which solutions are available and staff members with greater expertise on areas in which innovative solutions are required, and provide the opportunity to eliminate redundancy in the program work environment.

The Project Management Institute's *Standard for Program Management—Third Edition* (2013) discusses KM in section 6.3.3 but does not describe the importance of the KM plan. It notes the importance of organizing program knowledge for use as a reference and making it available to those who need it. The standard mentions three key elements of KM: collecting knowledge and sharing it across the program, using subject matter experts during the program, and using a program management information system for collecting program knowledge assets and artifacts. It further recognizes the importance of identifying, storing, and delivering key knowledge assets to stakeholders especially to support decision making.

Although the plan is prepared early in the program, it is an iterative document that should be reviewed and updated on a periodic basis.

Goals and objectives: This section lists the goals and objectives for KM in the program and describes how KM can contribute to the objectives of the program and the benefits to be realized. It may be that the program manager recognizes a need to quickly share information with his or her team, wishes to reduce the time required to locate information in documents and files, wants to reduce the time required to locate people with different areas of expertise on the team, or wants to avoid redundancy in the work that is done.

Roles and responsibilities: This section describes the key roles and responsibilities for KM in the program. A member of the program management team should have responsibility for KM activities, which would include preparing and updating the KM plan, ensuring that KM is a program package in the program work breakdown structure, determining the type of metadata tags to include on program artifacts, setting up a process so people can contribute knowledge assets or content to the knowledge repositor, approving knowledge assets or content for publication, publishing the content, prioritizing the content that is contributed, communicating the availability of content to stakeholders and members of the program management team, determining access rights to the content that is published, reviewing blogs and discussion forums for examples of content to share, establishing and using debriefing tools as team members and contractors leave the program, and sharing content within the team and the organization.

Examples of approaches to consider as metadata tags include: authored by or prepared by, date, document type (form, presentation, plan, report, register, meeting minutes, lessons learned sessions, program reviews, and stage gate reviews), modified by, description, key topic (if not evident by the type of document), component involved, stakeholders involved, benefits, and metrics.

This member of the program management team with principal responsibility for KM may work with someone in the program management office. He or she also should set up a process to work with at least one other person on the team to ensure that the content that is contributed is of high quality and should be published. This KM team member also can determine whether the content requires updates or should be deleted and should be responsible for knowledge transition to the program during the close program process.

Each member of the program management team, as well as customers and contractors, should identify knowledge assets or content and be able to contribute them.

Methods to communicate content availability within the program and within the organization include distribution lists, webinars, training sessions, discussion forums, blogs, podcasts, newsletters, white papers, collaboration rooms, document libraries, subscription notifications, and announcements.

Tools and techniques: This section describes the various tools and techniques to be used within the program team to share content that has been contributed. Examples of approaches include a program Web site, a shared drive, a knowledge repositor, a collaboration room, a document library, and discussion forums.

Training and orientation sessions: This section describes whether any training and orientation sessions may be needed if people working on the program are not accustomed to active involvement in KM. Examples of items to be included in these sessions are ways to share information, approaches to contribute knowledge assets or content, ownership of content and intellectual property issues, the need for nondisclosure issues, and security.

Metrics: This section states how the effectiveness of the KM process in the program will be evaluated. Examples of items to consider include the contribution on a regular basis of useful content, the "hit" rates per knowledge asset or content, the actual use of contributed content, and the ability to locate key content quickly or key subject matter experts easily.

Rewards and recognition: If the organization does not actively practice KM, this section may be required to state how best to recognize and reward those members of the program team who are active in the KM process. Incentives may be needed so people regularly contribute knowledge assets and content and apply them to their work on the program as appropriate.

Approvals: This section contains the written approval of the knowledge management plan by the program sponsor, program manager, program management office director, members of the Governance Board, and any other key stakeholders as appropriate.

Knowledge Management Plan Template

<Insert Program Name>
Knowledge Management Plan

Program name:	
Program manager:	PM's email address here as a hyperlink
Program sponsor:	
Actual start date:	
Approved end date:	
Program number:	
Revision history:	
Business unit:	

A. PURPOSE

A brief introductory statement defining the purpose of the knowledge management (KM) plan, such as:

> The knowledge management plan shows how knowledge management can improve the effectiveness and efficiency of overall program management, as it describes how people will be connected to one another in the program and to information generated throughout the course of the program.

_____

_____

_____

B. GOALS AND OBJECTIVES

This section lists the goals and objectives for knowledge management in the program and describes how KM can contribute to the program's objectives and benefits to be realized.

_____

_____

_____

C. ROLES AND RESPONSIBILITIES

This section describes the key roles and responsibilities for KM in the program.

_____

_____

_____

D. TOOLS AND TECHNIQUES

This section describes the various tools and techniques to be used by the program team to share content that has been contributed.

_____

_____

_____

E. TRAINING AND ORIENTATION SESSIONS

This section describes whether any training and orientation sessions may be needed if people working on the program are not accustomed to active involvement in KM.

_____

_____

_____

F. METRICS

This section states how the effectiveness of the KM process in the program will be evaluated.

_____

_____

_____

G. REWARDS AND RECOGNITION

This section describes how members of the program team may be recognized and rewarded for their KM contributions.

_____

_____

_____

H. APPROVALS

This section contains the approval of the knowledge management plan by the program sponsor, program manager, program management office, members of the Governance Board, and other key stakeholders as required.

SIGNATURES AND DATE APPROVAL OBTAINED

Program manager   _____

Program sponsor   _____

Program management office director _____

Governance Board chairperson _____

Governance Board member 1 _____

Governance Board member 2 _____

Governance Board member N _____

Stakeholder 1 _____

Stakeholder 2 _____

Stakeholder N _____

Governance Decision Register

The *Standard for Program Management*—Third Edition (2013) does not specify the format for recording governance decisions, but given the Governance Board is making key decisions throughout the program's life cycle, a decision register is a best practice to follow. Also, since many programs are complex and log, maintaining this register is useful as Board members may change during the program as well as possibly the program manager and members of his or her core team. Here, we present one proposed format.

Governance Decision Register Instructions

Purpose: The governance decision register is used throughout the program by the Governance Board to document the program decisions made at each meeting of the Board. It should be included with the minutes of the Board and distributed to stakeholders as described in the governance management plan and the program communications management plan. This register then is reviewed at subsequent meetings of the Governance Board. It is also useful to show progress in terms of meeting the program's objectives and to ensure they remain aligned with the organization's strategic objectives. Each decision documented in the register is used to help improve overall program results. Change requests may be required based on the type of decision and its impact on the program. These decisions, for example, may lead to a need to revise the governance management plan and the program management plan.

The governance decision register is helpful should members of the Governance Board change during the life of the program. The decision register should be part of the program's final records, as it can serve as an excellent source of lessons learned for future programs.

This governance decision register is designed as a table, and a description of its contents follows:

1. **Decision number:** Provide a number for the decision for tracking purposes.
2. **PWBS number:** Link the decision to a PWBS program package.
3. **Board meeting date:** State the date of the Governance Board meeting.
4. **Purpose of the Board meeting:** Describe the purpose of the Board meeting, such as a phase-gate review, overall program performance review, a meeting to initiate or terminate components, an overall review of the effectiveness of the governance management plan, or a request to close the program.
5. **Decision description:** Describe the actual decision and what is to be done. Provide sufficient detail for future use.
6. **Background information:** Provide background information as to why the decision was needed. Examples include escalated risk, issue, or audit recommendation by the program manager to the Governance Board; a requirement for approval to move to the next phase in the program's life cycle; to determine changes to ensure expected benefits are being realized; to determine whether new components should be added to the program or existing components terminated; to determine whether changes are required in program management so that best practices are followed; to ensure that the program's strategic objectives remain in alignment with the organization's strategic objectives; or to request a recommendation to close the program.
7. **Phase in the program management life cycle:** List the phase in the program management life cycle for this meeting of the Governance Board.
8. **Implementation date:** State the date the decision is to be implemented.
9. **Assigned to:** State the person who is responsible for implementing the decision and his or her contact information.
10. **Actual implementation date:** State the actual date the decision was implemented. Provide information if there is a variance between the actual date and the planned date.
11. **Notes:** Use this field for any additional notes about the decision.

Chapter 7

Program Life Cycle Management

Our days are so crowded and our hours are so few. And there's so little time and so much to do.

—Helen Steiner Rice

A program management life cycle can help program professionals in their quest to minimize their busy world, and it is during this life cycle in which the program's benefits are realized and sustained. However, the *Standard for Program Management*—Third Edition (2013) explains the program life cycle is not sequential as program components start and end at different times, and during the life cycle, the program produces 'a stream of deliverables' (p. 11). The life cycle spans the program's duration.

In the *Standard for Program Management*—Third Edition (2013), there is a Program Life Cycle Management Domain. It is divided into:

- Program Definition—program formulation and program preparation
- Program Benefits Delivery—component planning and authorization, component oversight and integration, and component transition and closure
- Program Closure—program transition and program closure

As noted, this life cycle serves to manage program activities as they are defined, as benefits are delivered, and as the program closes.

In PMI's *Examination Content Outline* (2011), the Program Life Cycle Domain is divided into:

- Initiating the Program
- Planning the Program
- Executing the Program
- Controlling the Program
- Closing the Program

Both approaches are similar as Initiating the Program relates to the Program Definition phase as it elaborates the program's strategic objectives and works toward program approval. Planning, Executing, and Controlling the Program relate to the Program Benefits Delivery phase, and Closing the Program relates to the Program Closure phase.

Because of the numerous possible plans, registers, and other templates, we have organized this chapter into five sub-chapters following the more familiar terms used in the *Examination Content Outline*.

Section 7A: Initiating the Program

> Without a charter, program execution is like flying without visual flight ranging in a cloud. Almost certainly in the heat of the moment you'll veer seriously toward somewhere you had no intention of going.
>
> **—Anonymous**

In his children's book, *Oh, the Places You'll Go!*, Dr. Seuss (1993) coined program management wisdom for the ages, warning eloquently of the scary proposition of making quick decisions "where the streets are not marked" and winding up in "weirdish wild space" and "useless places."

Having no use for useless places or weirdish wild space, in program initiation we produce a program charter for the program—building on the business case, program mandate, and the roadmap to guide us on the path to success. This program charter then links the program to the organization and its strategic goals and objectives. During this phase, we also may need to update the business case and the roadmap, in Chapter 3 in this book, as the program's financial framework is established, and the program charter is issued.

In the *Standard for Program Management*—Third Edition (2013), initiating involves defining the program, obtaining funding, and determining how the program will deliver its intended benefits. Other initiating activities include obtaining a sponsor; assigning a program manager; preparing estimates of scope, resources, and costs; performing an initial risk assessment; preparing the charter, and updating the business case and initial roadmap. It also suggests preparing the program management plan, which will be discussed in 7B.

A high-level overview of the initiating tasks from the Examination Content Outline (PMI, 2011) is as follows:

- Developing a program charter
- Translating high-level objectives into a scope statement (to be discussed in 7B)
- Developing a high-level milestone plan
- Developing a responsibility matrix
- Defining measurement criteria for project success in the program
- Conducting a kick-off meeting

Program Charter

Without a charter, the program manager has no authority to use resources and execute the program, and in fact, the program has no right to exist. The program charter is a major part of program initiation in the *Standard for Program Management*—Third Edition (2013) in section 8.3.1.6.

Program Charter Instructions

The program charter is a key program document. At the time the program is approved, it states the program vision, key objectives, expected benefits, constraints, and assumptions. It serves to link the program to the ongoing work of the organization. It also states the authority level of the program manager. When the program charter is approved, the program is officially authorized. A suggested format is:

Purpose: A brief introductory statement defining the purpose of the charter, such as:

> The program charter states the vision statement that defines the organization's end state for the program—a vital concept for successful program completion. It also states the program manager's authority and responsibility and formally authorizes the program.

Program vision: The program vision is the desired end state for the program. It describes why it will benefit the organization. It also describes the outcomes required to achieve the vision.

Justification: Each program has a key set of objectives to be accomplished. Objectives may involve the organization's business need for the program, a customer request, market demand, a regulation, etc. The program's objectives support the organization's business plan and strategic goals. The objectives should be measureable and state success criteria.

Strategic fit: The program's objectives are established in order to support those of the organization. This section lists the key strategic drivers for the program to show the link to the organization's strategic objectives and other ongoing strategic work in the organization.

Benefits: Programs are established in order to deliver benefits that may not be realized if projects and other work within the program were managed on an individual basis. Benefits enhance current capabilities or enable development of new capabilities. A benefit is an outcome of actions or behaviors to provide utility to the program's stakeholders. It is an improvement to the running of the organization, such as increased sales, reduced costs, or decreased waste. Benefits may be tangible (such as financial objectives) or intangible (such as customer satisfaction or improved employee morale). Benefits should be specific, measureable, actual, realistic, and time based. This section describes the benefits and how they are to be realized building on the initial list in Chapter 3, Program Strategic Management/Alignment.

Program constraints: Constraints are factors that limit the options of the program team. For example, a constraint may involve a regulatory requirement that must be met by a certain date. Constraints typically fall into categories such as time, cost, resources, or deliverables. Constraints are expected to change throughout the life of the program.

Program assumptions: This section describes the assumptions or any factors that are considered to be true, real, or certain in planning the program. Assumptions generally involve some degree of risk. They are expected to change throughout the life of the program.

Scope: This section describes what is within the scope of the program, and what is excluded from it. It serves as the basis to avoid possible scope creep from the beginning of the program.

Program components: This section states the various components to be part of the program to show how they will interface and interrelate throughout the program's life cycle in terms of a high-level plan for these components.

Known risks and issues: This section lists any risks and issues known at this time. They are expected to change during the life of the program.

Schedule: This section describes the planned length of the program and any milestones that are known at this time.

Resource requirements: This section lists required resources, both human and other resources, for the program and the cost of each resource for later in-depth planning.

Stakeholder considerations: This section builds on the initial stakeholder identification in Chapter 3, Program Strategic Management/Alignment, to list in more detail any key concerns of the stakeholders and the attitudes of the stakeholders who are expected to influence the program and be active in it at various phases of the program life cycle. It also includes a draft program communications management plan for these stakeholders since active communications are required at the time of program initiation.

Governance: This section describes the program's recommended governance structure as well as the governance structure for the various projects and non-project work that are part of the program, including reporting requirements. It also describes the program manager's level of authority.

Approvals: This section contains the approval of the program charter by the members of the Portfolio Review Board and other key stakeholders as required.

Program Charter Template

<Insert Program Name>
Program Charter

Program name:	
Program manager:	PM's email address here as a hyperlink
Program sponsor:	
Actual start date:	
Approved end date:	
Program number:	
Revision history:	
Business unit:	

A. PURPOSE

A brief introductory statement defining the purpose of the program charter, such as:

> The program charter states the vision statement that defines the organization's end state for the program to follow to successfully complete the program. It also states the program manager's authority and formally authorizes the program.

_____

_____

_____

B. PROGRAM VISION

This is a statement that describes the end state of the program and how it will benefit the organization. It also describes the outcomes required to achieve the vision.

_____

_____

_____

C. JUSTIFICATION

This section states the key objectives of the program as they support the business plan and strategic goals. It describes the objectives in measurable terms with specific success criteria.

_____

_____

_____

D. STRATEGIC FIT

This section states the how the program's objectives support the organization's strategic objectives. It also lists how the program supports other strategic initiatives in the organization.

_____

_____

_____

E. BENEFITS

This section states the expected benefits of the program to the organization and how they are planned to be realized.

_____

_____

_____

F. CONSTRAINTS

This section states the key constraints of the program. They are expected to change during the life of the program.

_____

_____

_____

G. ASSUMPTIONS

This section describes the assumptions that may affect the program. They are expected to change during the life of the program.

_____

_____

_____

H. SCOPE

This section describes what is within the scope of the program, and what is excluded from it.

_____

_____

_____

I. PROGRAM COMPONENTS

This section states the various components to be part of the program and includes a high-level plan for them.

_____

_____

_____

J. KNOWN RISKS AND ISSUES

This section lists any risks and issues known at this time.

_____

_____

_____

K. SCHEDULE

This section describes the planned length of the program and any milestones that are known at this time.

_____

_____

_____

L. RESOURCE REQUIREMENTS

This section lists required resources and the cost of each resource.

_____

_____

_____

M. STAKEHOLDER CONSIDERATIONS

This section lists any concerns of the key stakeholders who are expected to influence the program and be actively involved in it at various phases of the life cycle. It also includes a draft program communications management plan.

_____

_____

_____

N. GOVERNANCE

This section describes the program's governance structure and the governance structure planned for the program's components. It also describes the program manager's authority.

_____

_____

_____

O. APPROVALS

This section contains the approval of the program charter by the members of the Portfolio Review Board and any other key stakeholders as required.

SIGNATURES AND DATE APPROVAL OBTAINED

Portfolio Review Board chairperson _____

Portfolio Review Board member 1 _____

Portfolio Review Board member 2 _____

Portfolio Review Board member 3 _____

Stakeholder 1 _____

Stakeholder 2 _____

Stakeholder N _____

Program Financial Framework

The program financial framework is developed early in the program definition phase to set the stage for program financial matters. According the *Standard for Program Management*—Third Edition (2013), section 8.2.2, the framework defines available funding and outflows as well as financial constraints on income, spending and management. The information discovered in development of the program financial framework may have impact on the program business case. The financial framework then serves as a key document when the program's financial management plan is prepared, as discussed in Chapter 8B in this book.

Program Financial Framework Instructions

A suggested format is as follows:

Purpose: A brief introductory statement defining the purpose of the financial framework, such as:

> The program financial framework delineates the available funding sources and expected outlays as a basis for present and future budgeting to ensure adequate understanding of the special considerations involved in each funding source and expenditure destination. Additionally, it describes funding flows to ensure money is spent in the most efficient way.

Funding sources: While some programs only have a single funding source, others have multiple sources of funding. This section describes the various sources of funding for the program. Funding sources typically depend on the size, complexity, and type of program. Other factors include whether it is international or within a single country. The program may be funded internally and/or it may require external sources for funding.

Financial framework constraints: Each program has certain financial constraints. Examples of constraints that may be ones to consider include: how payment is to be made if on an international program; will a percent of contracts be held in retainage; since it is rare that a program, especially a long one, will receive all its funding at the beginning; when will funds be made available; whether government bonds will be used for funding that need to be created or sold; if funds are to be provided at pre-determined milestones;

whether compulsory employment will need to be included or if there is a requirement for compulsory use of a certain percent of indigenously produced products; and if there are any buy back arrangements that are required.

Financial framework assumptions: The program team will make assumptions concerning funding sources and funding goals for the program. As these assumptions may affect the program's deliverables, benefits, and objectives, they should be documented in this section and updated as changes occur during the program's life cycle.

Known risks and issues: Financial management, beginning with cost estimates and the financial framework, will have risks and issues to state. For example, a key risk could be if a contractor is subject to retainage, it may affect the contractor's ability to complete its portion of the program within budget. Another possible concern is if funding is to be provided at certain time period, and there is a budget cut that may occur. Such risks and issues are listed in this section.

Financial schedule: Given the length of programs, cash inflows tend to occur far more quickly than the program's benefits. A financial schedule is helpful to show both the inflows and outflows of funds for consideration by the Portfolio Review Board and other key stakeholders. As the schedule is prepared, if the program will be comprised of some existing components, review their schedules and infrastructure operational costs to date.

Resource requirements: This section describes key resource requirements from the program as the program manager will need to have a close working relationship with the financial organization and may need a core team member with a background in finance or support from the Enterprise Program Management Office.

Financial metrics: This section describes financial metrics to ensure that money is spent on the program in the most efficient way. These metrics should track to the business case, and as they are described in this document, it may be necessary to update the business case as a result.

Stakeholder and funding source considerations: This section states the key stakeholders with an influence over or an interest in the financial aspects of the program. It specifically lists those stakeholders that are expected to provide funding to the program and how best to engage them to ensure their support. It will serve as data to consider as the stakeholder engagement plan is prepared.

Approvals: This section contains the approval of the program financial framework by the members of the Portfolio Review Board and other key stakeholders as required.

The contents of this document can be refactored into a tabular format with accompanying discussion. A suggested document format is as follows.

Program Financial Framework Template

<Insert Program Name>
Program Financial Framework

Program name:	
Program manager:	PM's email address here as a hyperlink
Program sponsor:	
Actual start date:	
Approved end date:	
Program number:	
Revision history:	
Business unit:	

A. PURPOSE

A brief introductory statement defining the purpose of the program financial framework, such as:

> The program financial framework delineates the available funding sources and expected outlays as a basis for present and future budgeting and ensures adequate understanding of the special considerations involved in each funding source and expenditure destination. It describes funding flows to ensure money is spent in the most efficient way.

_____

_____

_____

B. FUNDING SOURCES

This section describes the financial sources and level of commitment for each one.

_____

_____

_____

C. FINANCIAL FRAMEWORK CONSTRAINTS

This section states the key constraints of the program from a financial viewpoint, such as certain funds must be used for certain purposes and not be used for other purposes. They may change during the life of the program.

_____

_____

_____

D. FINANCIAL FRAMEWORK ASSUMPTIONS

This section describes the assumptions that may affect the program's financial income, expenditures, management and outcomes.

_____

_____

_____

E. KNOWN RISKS AND ISSUES

This section lists any risks and issues known at this time, including but not limited to those potentially caused by the assumptions made.

_____

_____

_____

F. FINANCIAL SCHEDULE

In broad terms, this section describes the planned-for timing of income from financial sources and the timelines within which those funds are available for use. Consider component schedules and infrastructure and operational costs.

_____

_____

_____

G. RESOURCE REQUIREMENTS

This section lists required human and other resources planned for financial management and the cost of each resource.

_____

_____

_____

H. FINANCIAL METRICS

Metrics will be needed later in the program to correlate benefits to cost. Explain the expectedly relevant metrics and how they will be validated for use.

_____

_____

_____

I. STAKEHOLDER AND FUNDING SOURCE CONSIDERATIONS

Additional information about funding sources and stakeholders related to financial management may be vital. This section lists any concerns of the funding sources and key stakeholders and provides valuable input to the stakeholder engagement plan.

_____

_____

_____

J. APPROVALS

This section contains the approval of the program financial framework by the members of the Portfolio Review Board and any other key stakeholders as required.

SIGNATURES AND DATE APPROVAL OBTAINED

Portfolio Review Board chairperson _____

Portfolio Review Board member 1 _____

Portfolio Review Board member 2 _____

Portfolio Review Board member 3 _____

Stakeholder 1 _____

Stakeholder 2 _____

Stakeholder N _____

Section 7B: Planning the Program

> When you don't invest in infrastructure, you are going to pay sooner or later.
>
> **—Mike Parker**

In planning the program, selection of the projects and non-project work to be part of the program occurs. The high-level program road map is morphed into a detailed program road map. Here, the program charter is progressively elaborated to include much more detailed information, such as budget, schedule, and indeed a complete program management plan that describes the program's management infrastructure as prepared. That plan, by design, includes the tools and techniques that will eventually allow these "living documents" to evolve with developing realities.

As stated in the *Standard for Program Management*—Third Edition (2013), planning at the organizational level affects planning at the program level as the organization's strategic objectives continually evolve. Planning at the program level focuses on the program's deliverables and its benefits to ensure the benefits are realized, transitioned, and sustained. While a number of key plans (benefits realization, stakeholder engagement, communications management, change management, quality management, audit, knowledge management and governance) have been discussed in previous chapters, this chapter focuses on the program management plan and other subsidiary plans to best guide the program into executing, monitoring and controlling, closing, transitioning, and sustainment. Before the program can be closed officially, members of the Governance Board and the program sponsor will review the program management plan to ensure its contents have been fulfilled. Plans at the program level, however, do not replace plans at the project level but provide guidance to the project managers. In the *Standard for Program Management*—Third Edition (2013), the program management plan is part of program definition, see section 7.1.1.2 in the program definition phase. It is developed after the charter has been approved, and its approval by the Governance Board enables the program to move to the benefits delivery phase.

A high-level overview of the tasks In Planning the Program from PMI's *Examination Content Outline* follows:

- Developing a detailed scope statement
- Developing a Program Work Breakdown Structure
- Establishing the program management plan and schedule
- Optimizing the program management plan
- Defining a program management information system
- Identifying and managing unresolved project-level issues
- Developing a transition/integration/closure plan
- Developing key performance indicators
- Monitoring key human resources

Program Scope Management Plan

Just as the program scope statement defines the extents and boundaries within which to manage the program, the program scope management plan defines how to prepare the scope statement and the program work breakdown structure (PWBS) and how to manage, document, and communicate scope changes. It is noted the *Standard for Program Management*—Third Edition (2013) in section 8.9.1.

Program Scope Management Plan Instructions

The scope management plan will include the following:

Purpose: A brief introductory statement defining the purpose of the scope management plan, such as:

> The scope management plan describes the process of managing scope throughout the program.

This plan describes the process to follow to develop a detailed program scope statement, the PWBS, and the approach to follow to manage, document, and communicate scope changes. It also sets forth the relationship between product scope and program scope as it serves to identify the process and activities to produce deliverables and benefits to achieve program goals. This plan aligns the program's scope with the program's goals and objectives.

Product scope and program scope: This section states the difference between product scope and program scope. It recognizes that product scope involves the features and functions that characterize the products, services, and results of the program, while program scope describes the work required to deliver a major product, service, or benefit. Program scope also may be expressed as user stories or scenarios for some programs as described in the *Program Management Standard*—Third Edition (2013). Separate processes may be needed for product scope and program scope, which would be stated in this section. This section notes that completion of product scope is measured against program requirements, while completion of program scope is measured according to the program management plan and the benefits realization plan.

Expected stability of the scope of the program: This section describes the expected stability of the scope of the program based on information from program stakeholders. The stakeholder register, and its accompanying analysis, is reviewed and analyzed, and stakeholders are contacted to define and prioritize their requirements and to obtain inputs concerning program scope expectations. Program deliverable acceptance criteria are defined in this section for use in determining the overall stability of the scope of the program.

Scope statement development process: This section describes the process to follow to develop the scope statement for the program. A detailed program scope statement is critical to program success. The scope statement addresses the vision, range, capacity, and extent of the program endeavor. It serves to provide a common understanding of the program's scope by the stakeholders, as it establishes the direction to be taken and the essential aspects of the program to be accomplished. The process notes appropriate policies, procedures, and templates, which may be helpful in the development of the scope statement. It also describes tools and techniques to be used in the development of the scope statement, such as expert judgment, interviews, focus groups, and customer acceptance reviews.

Program work breakdown structure development process: This section describes the process to follow to develop the PWBS. The PWBS provides the framework for organizing and managing the work in the program. The objective is to structure the work in order that future activities and scope are easily managed. This section notes specific PWBS templates that may be considered, the use of expert judgment, lessons learned from previous programs, the various options to consider in program management in order that there is a balance between program deliverables and business objectives, the roles and level of involvement of specific individuals in the ownership of key activities, and methods to verify that the level of decomposition desired is necessary and sufficient.

Program scope change control process: This section describes the methods to use to identify program scope changes. The classification of these changes also is included in this section to determine the types of changes that the program manager can approve and others that need escalation to the program Governance Board. It also describes the process to follow to manage scope so scope creep does not occur, and states how program scope changes will be integrated into the overall program. Specific configuration management tools to be used are stated in this section to be able to monitor change control while enhancing organizational control so there is consistent documentation and product versions throughout the program. It also describes how to best communicate scope changes to program stakeholders.

Approvals: This section contains the written approval of the scope management plan by the program sponsor, program manager, program management office directors, members of the Governance Board, and other stakeholders.

Scope Management Plan Template

<Insert Program Name>
Scope Management Plan

Program name:	
Program manager:	PM's email address here as a hyperlink

Program sponsor:	
Actual start date:	
Approved end date:	
Program number:	
Revision history:	
Business unit:	

A. PURPOSE

A brief introductory statement defining the purpose of the scope management plan, such as:

The scope management plan describes the process of managing scope throughout the program.

_____

_____

_____

B. PRODUCT SCOPE AND PROGRAM SCOPE

This section states the difference between product scope and program scope. It describes how product scope and program scope are measured and states whether there is a need for separate processes to be used for product scope and program scope.

_____

_____

_____

C. EXPECTED STABILITY OF THE SCOPE OF THE PROGRAM

This section describes the expected stability of the scope of the program based on information from program stakeholders. It defines program deliverable acceptance criteria as part of this process.

_____

_____

_____

D. SCOPE STATEMENT DEVELOPMENT PROCESS

This section describes the process to follow to develop the scope statement for the program. It notes appropriate policies, procedures, and templates that may be helpful in the development of the scope statement and describes specific tools and techniques that will be used.

_____

_____

_____

E. PROGRAM WORK BREAKDOWN STRUCTURE DEVELOPMENT PROCESS

This section describes the process to follow to develop the PWBS. It notes the tools and techniques to be used, and the methods to verify that the level of decomposition desired is necessary and sufficient.

_____

_____

_____

F. PROGRAM SCOPE CHANGE CONTROL PROCESS

This section describes the methods to use to identify program scope changes. It includes how changes are classified and the process to follow to manage scope so scope creep does not occur. It states how program scope changes will be integrated into the overall program. It also describes specific configuration management tools to be used and how scope changes will be communicated.

_____

_____

_____

G. APPROVALS

This section contains the approval of the scope management plan by the program sponsor, program manager, program management office director, members of the Governance Board, and other key stakeholders.

SIGNATURES AND DATE APPROVAL OBTAINED

Program manager   _____

Program sponsor   _____

Program management office director _____

Governance Board chairperson _____

Governance Board member 1 _____

Governance Board member N _____

Stakeholder 1 _____

Stakeholder 2 _____

Stakeholder N _____

Program Scope Statement

Programs often fail because they do too many things in the heat of the battle that are being done by another team or organization, were not even part of the contract, or were not really needed for success. Ignoring this step can result in unnecessary requirements and serious cost overruns. The program scope statement is a living document for avoiding this type of failure and is used throughout the program life cycle. It is prepared in the program scope planning section (8.9.1) in the *Standard for Program Management*—Third Edition (2013).

Program Scope Statement Instructions

The program scope statement includes the following:

Purpose: A brief introductory statement defining the purpose of the scope statement, such as:

> The program scope statement defines the benefits, intent, and reason for pursuing a program by documenting what will be accomplished by the program.
>
> It serves as the basis for future program decisions and describes the scope boundaries of the program. It ensures the context and framework of the program are defined, assessed, and documented. It defines the direction to be taken in the program and the specific aspects to be accomplished. It is an iterative document that can be further refined as the program progresses through the life cycle.

Objectives and success criteria: This section includes the measurable success criteria of the program, including cost, schedule, technical, and quality objectives. Cost, schedule, and quality targets can be included. Benefits to be realized by the program should be mentioned with a reference to the benefits realization plan.

Program scope (inclusions and exclusions): This statement describes the characteristics of the product, service, or result that the program is undertaken to create. It is expected that these characteristics will be progressively elaborated as the project ensues.

Assumptions: This section lists and describes the specific program assumptions associated with the program scope, and the potential impact of the assumptions should they prove to be false.

Constraints: This section lists and describes the specific program constraints associated with the program scope that limit the team's options.

Deliverables and milestones: This section lists the deliverables, including both the outputs that comprise the product, service, or result, and the ancillary results, such as the program management reports and documentation. The milestones, proposed dates for those activities, and the deliverables are identified here, or a reference to the master program schedule may be used.

Acceptance criteria: This section identifies approval requirements that can be applied to items such as program objectives, deliverables (such as the manuals), documents (such as the component project plans and the performance tracking reports), and work products.

Risks: This section identifies the known scope risks to the program. It will be progressively elaborated during the course of the program.

Approvals: This section contains the written approval of the program scope statement by the program sponsor, program manager, program management office director, members of the Governance Board, and other stakeholders.

Program Scope Statement Template

<Insert Program Name>
Program Scope Statement

Program name:	
Program manager:	PM's email address here as a hyperlink
Program sponsor:	
Actual start date:	
Approved end date:	
Program number:	
Revision history:	
Business unit:	

A. PURPOSE

A brief introductory statement defining the purpose of the scope statement, such as:

The program scope statement defines the benefits, intent, and reason for pursuing a program by documenting what will be accomplished by the program.

_____

_____

_____

B. OBJECTIVES AND SUCCESS CRITERIA

This section includes the measurable success criteria of the program, including cost, schedule, technical, and quality objectives and achievement of program benefits.

_____

_____

_____

C. PROGRAM SCOPE (INCLUSIONS AND EXCLUSIONS)

This section describes the characteristics of the product, service, or result that the program is undertaken to create.

_____

_____

_____

D. ASSUMPTIONS

This section lists and describes the specific assumptions associated with the program scope, and the potential impact of the assumptions should they prove to be false.

_____

_____

_____

E. CONSTRAINTS

This section lists and describes the specific constraints associated with the program scope that limit the team's options.

_____

_____

_____

F. DELIVERABLES AND MILESTONES

This section is a statement of the deliverables, including both the outputs that comprise the product, service, or result, and the ancillary results, such as the program management reports and documentation.

_____

_____

_____

G. ACCEPTANCE CRITERIA

This section identifies approval requirements for items such as objectives, benefits realization plan, deliverables, documents, and work products.

_____

_____

_____

H. RISKS

This section identifies the known scope risks to the program.

_____

_____

_____

I. APPROVALS

This section contains the written approval of the program scope statement by the program sponsor, program manager, program management office director, members of the Governance Board, and other stakeholders.

SIGNATURES AND DATE APPROVAL OBTAINED

Program manager   _____

Program sponsor   _____

Program management office director   _____

Governance Board chairperson   _____

Governance Board Member 1   _____

Governance Board Member 2   _____

Governance Board Member N   _____

Stakeholder 1 _____

Stakeholder 2 _____

Stakeholder N _____

Program WBS

An abundance of references address the topic of work breakdown structure (WBS) generation, so we do not discuss this topic here. A salient point, however, is that the program WBS typically includes only one or two levels of each project's WBS— only enough to enable project tracking, monitoring, and control at the program level. The program WBS is discussed in section 8.9.1 in the *Standard for Program Management*—Third Edition (2013).

Program Resource Plan

While most resources and cost are probably at the component (project and non-project work) level and are thus managed there, the program level employs most or all of some resources for program management and governance. Necessary plans for efficiency and timeliness are in the program resource plan. In the *Standard for Program Management*—Third Edition (2013), its importance is stated in section 8.3.3.2, Program Infrastructure Development, noting that programs need people, tools, facilities, and finances specifically to manage the program, and it is described in more detail in section 8.6.1, Resource Planning.

Program Resource Plan Instructions

The program resource plan includes the following:

Purpose: A brief introductory statement defining the purpose of the program resource plan, such as:

> The program resource plan describes the resources required to manage the program. It includes methods to monitor and track resource use.

> These program resources include more than human resources and consist of materials, tools, equipment, facilities, and finances that will be used in the program. It also is important to recognize that this plan addresses resource use at the program level, not at the individual component level. Although the plan is drafted early in the program, it is an iterative document that should be further refined, especially since the preliminary requirements regarding the resources probably will change as the program progresses.

In the Project Management Institute's *Standard for Program Management—Third Edition* (2013), the program resource plan is discussed in section 8.6.1, Resource Planning in Program Resource Management. Required resources need to be available to program managers in a timely way to focus on benefit realization.

Internal and external required resources: This section describes specific internal and external, when appropriate, resources required for program success. It presents costs for each of these needed resources to assist in overall financial management and in determining the program's financial framework and financial management plan. It describes how some resources, such as office supplies are consumed by the program and managed as an expense, and emphasizes the need to account for all resources as well as those allocated appropriately at the component level. If external resources are required, the program manager prepares a statement of work to initiate the procurement process.

Organizational influences: This section defines various organizational influences that can impact the program team. It describes key procurement and personnel policies that may affect the acquisition of needed resources for the program. Also, it notes guidance on the appropriate use of program resources, such as depreciation or release of purchased equipment, sharing resources with other parts of the organization, or recurring costs of resources that are leased. It also states involvement with people from the quality department to ensure appropriate program quality assurance and quality controls are considered, including by contactors and components.

Methods to meet resource requirements: This section describes when and how resource requirements will be met in the program. It describes whether resources will be obtained internally or through procurements. It also describes the need to follow the issue escalation process should the program manager have difficulty acquiring needed resources to support the program.

Resource calendars: This section describes the necessary time frames for use of the program resources. These calendars can assist in development of the overall program schedule.

Tracking and monitoring: This section describes the methods to be used to track and monitor resource use throughout the program's life cycle. Since change is inevitable in all programs, the need for certain resources may change, especially as new components become part of the program, and other components are transitioned or terminated.

Resource prioritization: This focuses on scarce resources and the priority for use at the program and component level of these scarce resources. It describes the approaches the program manager will use in resource prioritization.

Approvals: This section contains the written approval of the program resource plan by the program sponsor, program manager, program management office director, members of the Governance Board, and other stakeholders.

Program Resource Plan Template

<Insert Program Name>
Program Resource Plan

Program name:	
Program manager:	PM's email address here as a hyperlink
Program sponsor:	
Actual start date:	
Approved end date:	
Program number:	
Revision history:	
Business unit:	

A. PURPOSE

A brief introductory statement defining the purpose of the resource management plan, such as:

The program resource plan describes the resources required to manage the program. It includes methods to prioritize, monitor, and track resource use.

_____

_____

_____

B. INTERNAL AND EXTERNAL REQUIRED RESOURCES

This section defines specific internal and external, when appropriate, resources required for program success. It also presents costs for each of these resources.

_____

_____

_____

C. ORGANIZATIONAL INFLUENCES

This section defines various organizational influences that can impact the program team's resource requirements, such as personnel and procurement policies and procedures, quality policies and standards, guidance regarding depreciation or release of equipment, sharing resources within the organization, and recurring costs of leased resources.

_____

_____

_____

D. METHODS TO MEET RESOURCE REQUIREMENTS

This section describes when and how resource requirements will be met in the program.

_____

_____

_____

E. RESOURCE CALENDARS

This section describes the needed time frames for use of the program resources.

_____

_____

_____

F. TRACKING AND MONITORING

This section describes the methods to be used to track and monitor resource use throughout the program's life cycle.

_____

_____

_____

G. RESOURCE PRIORITIZATION

This section describes the approaches the program manager will use to prioritize scarce resources across the program.

_____

_____

_____

H. APPROVALS

This section contains the approval of the program resource plan by the program sponsor, program manager, program management office director, members of the Governance Board, and other key stakeholders.

SIGNATURES AND DATE APPROVAL OBTAINED

Program manager _____

Program sponsor _____

Program management office director _____

Governance Board chairperson _____

Governance Board member 1 _____

Governance Board member 2 _____

Governance Board member N _____

Stakeholder 1 _____

Stakeholder 2 _____

Stakeholder N _____

Staffing Management Plan

Although this plan is not part of the *Standard for Program Management*—Second Edition (2008) or its Third Edition (2013), we believe it is an essential document for program management given the importance of acquiring, retaining, and releasing staff from the program.

Staffing Management Plan Instructions

The staffing management plan includes the following:

Purpose: A brief introductory statement defining the purpose of the staffing management plan, such as:

> The staffing management plan describes the timetable for staff acquisition and release from the program. It also includes identification of roles and responsibilities, plans for recognition and rewards, compliance considerations, and safety issues.

> It addresses availability of and competition for scarce human resources needed for the program. It designates roles and responsibilities required for program success. Although the plan is drafted early in the program, it is an iterative document that should be further refined, especially since the preliminary requirements regarding the people and competencies needed for the program are progressively elaborated as the program progresses.

Internal and external required resources: This section describes specific internal and external, when appropriate, human resources required for program success. Specific competencies are listed in this section. Competency refers to the skills and abilities required to complete program activities. The section lists the individuals and groups that may be part of the program organization and whether they are internal to the organization or external to it.

Organizational influences: This section defines various organizational influences that can impact the program team. It describes key personnel administration policies and procedures, standardized position descriptions, standardized performance plans, and the ability to acquire resources from groups external to the organization. It also specifically describes the assistance available to the program manager from the human resource department.

Program team roles and responsibilities: This section documents program team member roles and responsibilities so there is clarity of who is responsible for each program package in the program work breakdown structure. The objective is to ensure that the program team is aware of who is responsible for each of these program packages to ensure each one has an assigned owner, and duplicative work does not result. A role is defined as the portion of the program for which an individual is held accountable. Responsibility refers to the work the person is expected to perform in the program. A program organizational chart should be prepared and attached to the plan as a separate document. A resource assignment matrix also should be prepared.

Methods to meet human resource requirements: This section describes when and how human resource requirements will be met in the program. It describes how staff members will be acquired and whether or not the team will be virtual or collocated.

Resource calendars: This section describes the necessary time frames for members of the program team to join the program and begin their work, as well as when recruiting or acquisition activities will begin. A histogram of human resource requirements may be attached to this plan.

Staffing release: This section describes the method and timing of releasing staff members from the program team. Its purpose is to provide a smooth transition as staff members are released. Methods for identifying lessons learned for future program improvements can be part of this section of the plan as individuals move on to other assignments. Such a smooth transition can improve overall team morale.

Rewards and recognition: This section presents clear criteria for rewards and a planned system for their use on the program. Various types of available rewards are included in this section. The purpose of this section is to ensure there is a set timetable so recognition is not forgotten as program activities are completed.

Compliance requirements: This section describes strategies for compliance with various regulations, organizational policies, and human resource policies to be followed in the program. It also includes any safety policies and procedures.

Approvals: This section contains the written approval of the staffing management plan by the program sponsor, program manager, program management office director, members of the Governance Board, and other stakeholders.

Staffing Management Plan Template

<Insert Program Name>
Staffing Management Plan

Program name:	
Program manager:	PM's email address here as a hyperlink
Program sponsor:	
Actual start date:	
Approved end date:	
Program number:	
Revision history:	
Business unit:	

A. PURPOSE

A brief introductory statement defining the purpose of the staffing management plan, such as:

> The staffing management plan describes the timetable for staff acquisition and release from the program. It also includes identification of roles and responsibilities, plans for recognition and rewards, compliance considerations, and safety issues.

_____

_____

_____

B. INTERNAL AND EXTERNAL REQUIRED RESOURCES

This section defines specific internal and external, when appropriate, human resources required for program success. It lists specific competencies that are needed.

_____

_____

C. ORGANIZATIONAL INFLUENCES

This section defines various organizational influences that can impact the program team, such as personnel administration policies and procedures, standard position descriptions, standard performance plans, and the ability to acquire resources from external units.

_____

_____

_____

D. PROGRAM TEAM ROLES AND RESPONSIBILITIES

This section documents program team member roles and responsibilities for each program package in the program work breakdown structure. A program organization chart should be attached to this plan as a separate document. A resource assignment matrix also may be included.

_____

_____

_____

E. METHODS TO MEET HUMAN RESOURCE REQUIREMENTS

This section describes when and how human resource requirements will be met in the program. It states how staff members will be acquired and whether the team will be virtual or collocated.

_____

_____

_____

F. RESOURCE CALENDARS

This section describes the necessary time frames for members of the program team to join the program and begin their work. It also states when recruiting or acquisition activities will begin. A histogram may be attached to this plan.

_____

_____

_____

G. STAFFING RELEASE

This section describes the method and timing of releasing staff members from the program team so there is a smooth transition. It also includes methods to identify lessons learned as staff members move on to other assignments.

_____

_____

_____

H. REWARDS AND RECOGNITION

This section presents the criteria for rewards and a planned system for their use in the program. It helps to ensure there is a set timetable so recognition is not forgotten as program activities are completed.

_____

_____

_____

I. COMPLIANCE REQUIREMENTS

This section describes strategies for compliance with various regulations, organizational policies, human resource policies, and safety policies and approvals.

_____

_____

_____

J. APPROVALS

This section contains the approval of the staffing management plan by the program sponsor, program manager, program management office director, members of the Governance Board, and other key stakeholders.

SIGNATURES AND DATE APPROVAL OBTAINED

Program manager _____

Program sponsor _____

Program management office director _____

Governance Board chairperson _____

Governance Board member 1 _____

Governance Board member 2 _____

Governance Board member N _____

Stakeholder 1 _____

Stakeholder 2 _____

Stakeholder N _____

Program Requirements Document

Requirements documents have been around for many years. We chose not to cover this document in detail, except to emphasize the fact that both program requirements and component requirements must be included, where program requirements are much higher in level. Component requirements, on the other hand, are written only to such a level that the component can begin work. A requirements document is not discussed separately in the *Standard for Program Management—Third Edition* (2013); however, the importance of requirements in a variety of areas in program management is mentioned frequently.

Program Requirements Document Instructions

The program requirements document includes the following:

Purpose: A brief introductory statement defining the purpose of the program requirements document, such as:

The program requirements document describes the high-level requirements to ensure that the program delivers its expected benefits.

The program requirements document serves as the basis for the component requirements document, which is used to develop the technical requirements for each of the component's deliverables. It identifies and details program specifications and specific outcomes to be achieved by the program during implementation.

The program requirements document should be prepared once program goals and objectives are defined. It builds on the requirements listed in the business case, the scope statement, and the program roadmap. Programs must ensure that a detailed process is followed as this document is developed, since it is imperative for successful program implementation.

Requirements management process: This section describes how the project manager and his or her team will develop the program requirements in sufficient detail to ensure they address all program components. It describes roles and responsibilities of program management team members and states the level of detail to be provided in the program requirements.

Requirements gathering tools: This section describes the specific requirements gathering tools the program management team will use as it develops the program requirements. Examples include the use of brainstorming, interviews, focus groups, questionnaires, and surveys as the program management team actively engages its stakeholders in this process.

Requirements analysis process: This section then describes the approach the program management team will follow to ensure that the requirements it has identified are complete. It describes the decomposition process the team will follow to take the high-level requirements into ones that can be further detailed by each component.

Requirements review process: This section states how the program management team will review the requirements that have been developed to ensure they are consistent, accurate, and complete. It describes specific approaches that may be used, such as various reviews, both informal and formal. It notes how the team will verify that best practices were followed and describes whether subject matter experts will be used in the reviews. It also states the level of customer involvement and that of the senior managers in the organization.

Requirements validation and verification: This section describes methods to be used to validate and verify the requirements that have been developed in order to ensure that they support the program's business case and have been decomposed to the level required for implementation. It then describes how the products of the program will be verified against the requirements and the acceptance process that will be followed both at a detailed and at a system level. It states the extent of any testing to be used, inspections to be conducted, or other approaches that will be followed.

Approvals: This section contains the written approval of the requirements document by the program sponsor, program manager, program management office director, members of the Governance Board, and any other key stakeholders as appropriate.

Program Requirements Document Template

<Insert Program Name>
Requirements Document

Program name:	
Program manager:	PM's email address here as a hyperlink
Program sponsor:	
Actual start date:	
Approved end date:	
Program number:	
Revision history:	
Business unit:	

A. PURPOSE

A brief introductory statement defining the purpose of the requirements document, such as:

> The program requirements document describes the high-level requirements to ensure that the program delivers its expected benefits.

_____

_____

_____

B. REQUIREMENTS MANAGEMENT PROCESS

This section describes the process the program management team will follow to develop the program requirements in sufficient detail to ensure all program components are addressed. It describes team member roles and responsibilities and states the level of detail to be provided in the program requirements.

_____

_____

_____

C. REQUIREMENTS GATHERING TOOLS

This section describes the specific requirements gathering tools the program management team will use.

_____

_____

_____

D. REQUIREMENTS ANALYSIS PROCESS

This section describes the approach of the program management team to ensure the requirements it has identified are complete.

_____

_____

_____

E. REQUIREMENTS REVIEW PROCESS

This section states how the program management team will review the requirements to ensure they are consistent, complete, and accurate. It describes specific reviews that will be held, attendees at these reviews, and their roles.

_____

_____

_____

F. REQUIREMENTS VALIDATION AND VERIFICATION

This section describes methods to validate and verify the requirements to ensure they support the program's business case and have been decomposed to the necessary level. It states how products will be verified against the requirements and the acceptance process to follow.

_____

_____

G. APPROVALS

This section contains the approval of the program requirements document by the program sponsor, program manager, program management office director, members of the Governance Board, and other key stakeholders as required.

SIGNATURES AND DATE APPROVAL OBTAINED

Program manager _____

Program sponsor _____

Program management office director _____

Governance Board chairperson _____

Governance Board member 1 _____

Governance Board member 2 _____

Governance Board member N _____

Stakeholder 1 _____

Stakeholder 2 _____

Stakeholder N _____

Risk Management Planning Meeting Agenda

Documents such as the risk management plan and more are generated at risk management planning meetings. This agenda should be of significant help in assuring success in this critical endeavor. This meeting is not discussed as a way to prepare the risk management plan for the program in section 8.7.1 in the *Standard for Program Management*—Third Edition (2013); however to encourage stakeholder engagement in the risk planning process, we feel it is a best practice to hold this meeting to gather perspectives as the plan is prepared.

Risk Management Planning Meeting Agenda Instructions

All programs will have risks, and risk management planning is an important component to continually assess the program for risks that are opportunities to pursue as well as possible threats to the program. Holding a risk management planning meeting with a variety of program stakeholders can facilitate their engagement in the program's risk management activities especially as the plan is prepared.

Typical participants include the program manager, members of the program management team, a risk management specialist in the organization, the director of the program management office, and the portfolio manager. The program manager should review the stakeholders that have been identified to date and the stakeholder engagement plan and decide whether or not to invite other stakeholders to the meeting.

Prior to the meeting, participants should receive copies of the program mandate, business case, stakeholder engagement plan, program charter, program roadmap, scope statement, program work breakdown structure, governance plan, and resource plan. The program manager should consult the organization's knowledge management repository for any lessons learned from previous or ongoing projects or programs that may be helpful to this program concerning risk management. If component projects have been under way before inclusion in the program and have prepared a risk management plan, the program manager should provide them to attendees. If there are existing templates in the organization for use in risk management, participants should receive them prior to the planning meeting.

An agenda for the meetings is as follows:

PARTICIPANTS (NAMES/ORGANIZATION)

Date: _____

Time: _____

Place: _____

Program overview: The program manager presents a brief high-level overview of the program as to why the program was undertaken and where it stands in the program management life cycle. If other meetings are held at a later time to refine or update the plan, he or she should provide a short summary of the last meeting held and should provide participants with the latest program risk register.

Program risk management activities: The risk management planning meeting defines the risk management activities to be followed in the program. During this meeting, participants determine the following:

    Specific risk categories that are expected to affect the program

    A risk breakdown structure

    Standard definitions for risk levels, probability, and impact to be followed throughout the program

    A probability/impact matrix

    Risk identification checklists

A method to follow to assess component-level risks for their effect on the program and to see if there are any interdependencies

The use of modeling techniques

The need for the use of independent experts to periodically review risk management activities

Items to be included in the risk management plan

Items to be included in the risk register

Risk management responsibilities: There are a number of responsibilities associated with risk management, which include:

Authority level for decisions regarding risks that affect the program

Maintenance of the risk register

Updates to the risk management plan

Preparation of agendas and minutes from risk reviews

Use of risk management software

Including risk items on the program's schedule and updating them as needed

Including risk as part of the program's budget and updating requirements as needed

Program risk management approach: This meeting addresses items to be included in the risk management plan and an overall approach for risk management for the program.

Decisions: This section documents the decisions made during this meeting.

Signatures of Participants

Program manager _____

Participant 1 _____

Participant 2 _____

Participant N _____

Program Risk Management Plan

Programs must have a disciplined way to deal with risks. This plan, while a living document throughout the program, would normally remain somewhat static, defining the processes that populate and manage the risk register, generate risk response plans, contingency plans, etc. It is discussed in section 8.7.1, Program Risk Management Planning, in the *Standard for Program Management*—Third Edition (2013).

Program Risk Management Plan Instructions

The program risk management plan includes the following:

Purpose: A brief introductory statement defining the purpose of the risk management plan, such as:

The risk management plan describes the process of determining how to approach, plan, and execute the program risk activities.

This plan recognizes that risks are a series of events that, if they occur, may have a positive or a negative impact on the program. At the program level, risks tend to occur from external factors or from individual components. If they are due to risks at the component level, the program management team must assess them for impact to other components or to the entire program. The program management team must ensure that the level, type, and visibility of risks are appropriate to the specific program since some programs may require greater attention to risk management than others, especially programs that are new in numerous aspects to the organization, are considered complex, and involve new technologies and systems.

Risk management approach: This section describes how the program management team will handle risk management in the program. The organization's culture risk profiles, and thresholds, and market factors affecting the program are described as appropriate. It states the methodology to be followed, tools and techniques to be used, and any other data sources, such as a knowledge repository or lessons learned database, that may be helpful. If an organization-wide risk management approach, such as pre-defined risk categories, roles and responsibilities, common terms to use, templates, risk statement formats, and authority levels for decisions, is in place, it notes whether or not it will be followed or whether changes to it are warranted. It also describes how proposed changes will be identified and managed in order that they do not pose additional risks to the program. It notes the process for the program manager to learn of changes at the project level that may affect other projects in the program or the overall program in order that the program management team can then review these changes for possible new risks.

Program risk categories: This section describes specific categories of risks that may affect the program. Typically these categories may include external environmental risks, such as new regulations or standards or strategic issues; program-level risks, such as ones involving program stakeholders, the Governance Board, and how the program components are organized; project-level risks, while managed by the project manager, that may affect other components in the program; operational-level risks, such as ones that involve transition of the program benefits into the overall operations of the organization or to the customer or possible new systems in the organization that may affect program benefits; portfolio-related risks, since the portfolio level defines the organization's strategic intent, which then may affect the overall priority of the program in terms of resources; and benefit-related risks, which involve the impact of risks from the component level to the overall delivery of program benefits. If a risk breakdown structure is prepared, it should be attached to this plan.

Roles and responsibilities: This section describes the roles and responsibilities of the program management team regarding risk management planning, identification, analysis, response planning, monitoring, and control. It may be appropriate on certain projects to designate a member of the core team to focus on risk management at the program level and to work with the various project managers to review risks at the project level and to conduct project risk audits. This section describes how risks will be escalated from component managers to the program manager, and from the program manager, as appropriate, to the Governance Board, and to describe specific roles and responsibilities at each level. It also addresses roles and responsibilities for risk management of inter-component risks and for analysis of root causes of these risks in order that an appropriate risk response can be provided.

Probability and impact matrix: This section describes definitions of probability and impact for use on the program. These definitions can be related to the program's objectives or to the program's expected benefits. Numeric and non-numeric approaches can be used. This matrix then serves as a table in which to show specific combinations of risk probability and risk impact that are then considered to be high, medium, or low in terms of importance to help plan appropriate risk responses. This matrix serves to help prioritize those risks that require the greatest attention from the program management team.

Risk analysis considerations: In order to support an environment that is conducive to risk management, this section states how the program team assists in risk analysis. To do so, this section describes available information, available resources, time and cost, the quality of the available information, and ways to be apprised of activities that are outside of the direct control of the project teams such as establishing different communication channels between program components and with other programs.

Risk management budget: This section describes the budget or contingency that will be set aside to focus on risk management throughout the program life cycle. It therefore provides a cost estimate for risk management at the program level to be used to determine the funding and resources required. It also describes the process to follow when it is necessary to allocate contingency reserve in order to respond to program-level risks.

Risk management schedule: Although risk management is a continual activity throughout the program, this section describes the specific risk management activities that should be included in the program's schedule. Items such as when a program risk management planning meeting, program risk reviews or audits, and analysis of lessons learned from risk management initiatives in the program are planned are examples of activities to be part of the schedule.

Stakeholder tolerances for risk: This section describes the tolerance level of the key program stakeholders in terms of risk. The stakeholder register and the stakeholder engagement plan can be used as this section is prepared, and interviews can be conducted with key stakeholders. The purpose is to show those stakeholders in terms of their influence on the program in generating and responding to program risk based on their tolerance for risk. It considers the culture of the organization in terms of its approach to risk management.

Risk reporting: This section describes the content and format for the program's risk register, which will be used throughout the program for risk reporting and to assist in risk identification, analysis, response planning, and risk monitoring and control. It also describes communication approaches to be used in the program with the various stakeholders and the Governance Board members concerning risks.

Risk tracking: This section describes the process to track identified risks and to recognize any new risks that may affect the program. It also describes how the program's risk management process will be audited and the frequency of the audits to be conducted. It describes the process for documenting lessons learned based on the program's risk management activities.

Approvals: This section contains the written approval of the risk management plan by the program sponsor, program manager, program management office director, members of the Governance Board, and other stakeholders.

Program Risk Management Plan Template

<Insert Program Name>
Risk Management Plan

Program name:	
Program manager:	PM's email address here as a hyperlink
Program sponsor:	
Actual start date:	
Approved end date:	
Program number:	
Revision history:	
Business unit:	

A. PURPOSE

A brief introductory statement defining the purpose of the risk management plan, such as:

The risk management plan describes the process of determining how to approach, plan, and execute the program risk activities.

_____

_____

_____

B. RISK MANAGEMENT APPROACH

This section describes how the program management team will handle risk management in the program. It states the methodology to be followed, tools and techniques to be used, and any other data sources that may be helpful. It describes how proposed changes will be identified and managed in order that they do not pose additional risks to the program.

_____

_____

_____

C. PROGRAM RISK CATEGORIES

This section describes specific categories of risks that may affect the program. If a risk breakdown structure is prepared, it should be attached to this section.

_____

_____

_____

D. ROLES AND RESPONSIBILITIES

This section describes the roles and responsibilities of the program management team regarding risk management planning, identification, analysis, response planning, abd monitoring and control. It also describes how risks will be escalated from component managers to the program manager, and from the program manager, as appropriate, to the Governance Board, and to describe roles and responsibilities at each level. It shows roles and responsibilities for inter-component risks.

_____

_____

_____

E. PROBABILITY/IMPACT MATRIX

This section defines probability and impact in terms of the program.

_____

_____

_____

F. RISK ANALYSIS COSIDERATIONS

This section describes the methods the program management team can use to assist in risk analysis to best provide an environment conducive to risk management.

_____

_____

_____

_____

G. RISK MANAGEMENT BUDGET

This section describes the process to be followed to prepare a risk management cost estimate that then will be used for a risk management budget or contingency throughout the program life cycle. It describes the process to follow when it is necessary to allocate contingency reserve in order to respond to program-level risks.

_____

_____

_____

H. RISK MANAGEMENT SCHEDULE

This section describes the specific risk management activities that should be included in the program's schedule.

_____

_____

_____

I. STAKEHOLDER TOLERANCES FOR RISK

This section describes the tolerance level of the program stakeholders in terms of risk to show those stakeholders based on their level of influence on the program in generating and responding to program risk. It also considers the organization's culture regarding risk management.

_____

_____

_____

J. RISK REPORTING

This section describes the content and format for the program's risk register. It also describes communication approaches to be used in the program with the various stakeholders and the Governance Board concerning risks.

_____

_____

_____

K. RISK TRACKING

This section describes the process to follow to track identified risks and to recognize any new risks that may affect the program. It also describes how the program's risk management process will be audited and the frequency of the audits, as well as the process to document lessons learned based on the program's risk management activities.

_____

_____

_____

L. APPROVALS

This section contains the approval of the risk management plan by the program sponsor, program manager, program management office director, members of the Governance Board, and other key stakeholders.

SIGNATURES AND DATE APPROVAL OBTAINED

Program manager _____

Program sponsor _____

Program management office director _____

Governance Board chairperson _____

Governance Board member 1 _____

Governance Board member 2 _____

Governance Board member N _____

Stakeholder 1 _____

Stakeholder 2 _____

Stakeholder N _____

Program Risk Register

The risk register is a critical program decision-making tool. It combines in one place the needed decision criteria such as the likelihood, status, and impact of program risks. It is first discussed in section 8.7.2 in the *Standard for Program Management— Third Edition* (2013).

Program Risk Register Instructions

Purpose: The risk register is used throughout the program. Its use begins when a risk is identified and continues until all risks have been closed or are no longer considered a threat or opportunity to the program. It complements the program risk management plan but does not replace it.

With continuous use of the risk register throughout the program, and monitoring the impact and probability of occurrence of a program risk, the program management team can monitor possible risks and use proactive actions to prevent their occurrence. Such an approach can be effective in ensuring the risk responses have the desired effect.

Resolution of certain risks may be handled by the risk owner; others may need to be escalated to the program manager or to the Governance Board. The decision as to which risk response to use and when to close a risk may require involvement from other stakeholders. Change requests to various items in the program may lead to a need to revise the risk register or to add newly identified risks to it. Even though the program has an issue register, it does not replace the need for a risk register.

In the Project Management Institute's *Standard for Program Management—Third Edition* (2013), the risk register updated as described in section 8.7.3, Program Risk Analysis; 8.7.4 Program Risk Response Planning; and 8.7.4, Program Risk Monitoring and Control.

The risk register is designed as a table, and a description of its contents follows:

1. **Risk identification number:** Assign an identification number to the risk.
2. **PWBS number:** Link the risk to the corresponding program package in the program work breakdown structure.
3. **Risk description:** Describe the risk and why it is significant to the program.
4. **Identified by:** List the person who identified the risk and provide contact information.
5. **Date:** State the date the risk was identified and added to the register.
6. **Risk type:** Describe the type of risk. Follow the categories established in the risk management plan.
7. **Root cause:** Describe the program-level root cause of the risk in order to determine potential responses. Doing so can also help group risks into causes for analysis and review and avoid potential future program risks.
8. **Probability/impact:** State the probability/impact of the risk to the program. Use a probability/impact matrix.
9. **Risk owner:** Identify a member of the program management team to be responsible for the risk and to track it until it is resolved and closed. Provide contact information for the owner.
10. **Symptoms/warning signs:** Identify various triggers or warning symptoms to alert the program management team that the risk might occur. For example, poor morale on the program management team might indicate that the program's resources may be affected as people may choose to leave the program for other opportunities.

11. **Proposed response:** State the proposed risk response. Examples include:
    - Negative risks or threats:
      - Avoid
      - Transfer
      - Mitigate
      - Accept
    - Positive risks or opportunities:
      - Enhance
      - Exploit
      - Share
      - Accept

    Note that if a risk is accepted, a contingency plan will be required. Follow the process outlined in the risk management plan in the risk management budget section in allocating contingency reserve to respond to program-level risks. Describe the budget and schedule activities needed to implement the response. Follow the template for the contingency plan.

12. **Actual response:** If the risk does in fact occur, indicate which response strategy was selected. If it was not the planned response, describe why for purposes of lessons learned.

13. **Approved by:** State who approved the risk response. Examples include:
    a. Risk owner
    b. Program manager
    c. Governance Board

14. **Date:** State the date the risk response was approved.

15. **Subsequent impacts:** State any subsequent impacts as a result of the risk response. Examples include the need to use a fallback plan if the response was not effective, residual risks that remain and may affect the program at a later time after the response was implemented, and the identification of subsidiary risks as a result of implementing the response. Use the risk register to state any residual or secondary risks as appropriate.

16. **Date closed:** State the date the risk was closed.

17. **Notes:** Use this field for any additional notes about the risk.

Probability/Impact Matrix

The probability/impact matrix is used to determine the overall priority of each risk in the risk register to the program. It also serves as way to communicate the importance of the risk to program stakeholders, as a red, yellow, green approach may be helpful as part of this matrix.

Risk probability describes the degree of uncertainty, while risk impact states the magnitude of the effect the risk will have on program objectives. The example in the table presents one approach to consider. It also relates to the variances based on stakeholder tolerances for risks as described in the risk management plan.

Probability/Impact Matrix

Impact	Probability				
	Very High	*High*	*Medium*	*Low*	*Very Low*
Catastrophic	Very High	High	Moderate	Moderate	Very Low
Critical	High	High	Moderate	Very Low	Very Low
Marginal	Moderate	Moderate	Very Low	Very Low	None
Negligible	Moderate	Low	Very Low	None	None

In preparing the matrix, consider items such as:

The time the risk is expected to occur

Whether the risk is expected to occur at a certain phase in the program's life cycle

Whether the probability or the impact is expected to change over time

Whether there is a date after which the risk is expected not to affect the program

The ease of managing and controlling the risk

The completeness of the description of the risk in the risk register

The level of vulnerability to the project should the risk occur

Whether the cause of the risk is due to the program or to other events

Update and review the probability/impact matrix as risk responses are implemented, risks are closed, and new risks are added to the register. Note the results of the probability/impact matrix in the risk register.

Contingency Plan

The most serious program risks need at least a preliminary plan that can be implemented should they occur. Not having a contingency plan for an occurring risk is serious and possibly a threat to overall program success. The contingency plan is discussed in section 8.7.4, Program Risk Response Planning the *Standard for Program Management*—Third Edition (2013). This section also mentions contingency budgets and contingency reserves, pointing out contingency reserves are for use at the program level as each project also should have a contingency reserve. Cost and schedule contingency reserves are also mentioned in section 8.7.5, Program Risk Monitoring and Control.

Contingency Plan Instructions

The contingency plan will include the following:

Purpose: A brief introductory statement defining the purpose of the contingency plan, such as:

The contingency plan describes the impact on the program for those risks in the risk register that require such a plan.

This plan complements the program risk management plan and the risk register. It is prepared only for those risks in which a contingency plan is warranted. It is used for unknown risks that cannot be managed proactively and also for those identified risks in which the response strategy is to actively accept the risk. Time, money, and resources then will be required if this is the case, regardless if the risk is a threat to the program or an opportunity. The program management team uses the contingency plans to assess impacts across the program and to see if any of them are interrelated to mitigate risk activities that could affect the program components.

When a contingency plan is implemented, it will require a change request and should be handled as part of the program's integrated change control system.

Trigger conditions: Each risk should have certain triggers or warning signs associated with it. This section lists these triggers, which then will result in executing the contingency plan, such as missing a key milestone.

Cost impact: Each contingency plan requires a cost estimate. This section states the needed budget to implement the contingency plan and considers the associated impacts on the schedule, resources, and deliverables as the estimate is prepared.

Schedule impact: If there is a need to implement a contingency plan, it will affect the program's schedule. This section describes this impact and shows the need to update or change duration estimates of key activities or to change or add milestones as appropriate.

Required resources: If a contingency plan is executed, resources will be required. This section states these necessary resources.

Responsibilities: This section describes the responsibilities associated with implementing the contingency plan. It describes the various stakeholders and their level of involvement. It states the primary point of contact on the program management team.

Approvals: This section contains the written approval of the contingency plan by the program manager and other stakeholders as required.

Contingency Plan Template

<Insert Program Name>
Contingency Plan

Program name:	
Program manager:	PM's email address here as a hyperlink
Program sponsor:	
Actual start date:	
Approved end date:	

Program number:	
Last updated:	
Business unit:	

A. PURPOSE

A brief introductory statement defining the purpose of the contingency plan, such as:

> The contingency plan describes the impact on the program for those risks in the risk register that require such a plan.

_____

_____

_____

B. TRIGGER CONDITIONS

The risk register lists those risks for which a contingency plan will be needed. This section describes the triggers or warning symptoms associated with these risks.

_____

_____

_____

C. COST IMPACT

This section describes the cost estimate if the contingency plan is to be implemented.

_____

_____

_____

D. SCHEDULE IMPACT

This section describes the effect on the schedule if the contingency plan is implemented, as duration estimates may require change, and milestones may need to be added or deleted.

_____

_____

_____

E. REQUIRED RESOURCES

This section states the necessary resources to implement the contingency plan.

_____

_____

_____

F. RESPONSIBILITIES

This section describes the responsibilities associated with implementing the contingency plan and states the primary point of contact on the program management team.

_____

_____

_____

G. APPROVALS

This section contains the written approval of the contingency plan by the program manager and other stakeholders as required.

SIGNATURES AND DATE APPROVAL OBTAINED

Program manager   _____

Stakeholder 1   _____

Stakeholder 2   _____

Stakeholder N   _____

Schedule Management Plan

The program must have discipline and control in maintenance of its official schedule. The schedule management plan describes how this discipline will be achieved and is prepared as noted in section 8.8.1 Program Schedule Planning in the *Standard for Program Management*—Third Edition (2013).

Schedule Management Plan Instructions

The schedule management plan will include the following:

Purpose: A brief introductory statement defining the purpose of the schedule management plan, such as:

> The schedule management plan identifies an agreed-upon sequence of component deliveries (projects and non-project work) to enable

them to be planned and managed. It sets forth a scheduling methodology, states the scheduling tool to be used, and sets the format and criteria for monitoring and controlling the program schedule.

This plan provides the program management team and stakeholders with an approach to how the schedule will be managed throughout the program. It enables the program manager to identify any risks associated with the schedule and any component issues that may need to be escalated for resolution. A common understanding among program stakeholders is required as to how to clearly define the program's schedule status. Although the plan is drafted early in the program with a high-level master schedule to show the benefits and outputs from each of the components, it is an iterative document that should be further refined in each phase of your program.

In the Project Management Institute's *Standard for Program Management—Third Edition* (2013), the schedule management plan is discussed in section 8.8.1, Program Schedule Planning. The emphasis is on the order and timing of the program's components to ensure the program's benefits are realized as planned. It builds on the road map, the scope management plan, and the PWBS. As it is developed, it then may require updates to the roadmap and inputs to the risk register.

Program schedule constraints: This section describes constraints, both internal and external, that may affect the program schedule. Examples include funding, resource availability, technical considerations, contracts, predetermined milestones, environmental issues, other external dependencies, and other factors in the program environment.

Program schedule standards: Each organization has its own schedule standards. This section describes the standards applicable to the program such as data formatting, versioning, the methodology to use, performance thresholds, and the content of presentations and reports. This section also notes any processes in place for schedule updates, the need for training for the program team, and considerations concerning contractors and their access to the schedule.

Delivery of incremental benefits: Since the purpose of a program is to deliver benefits that might not be realized if the projects in the program were managed individually, this section of the plan provides a description of the order in which the incremental benefits should be delivered. It also contains component milestones to show when program benefits are expected to be realized.

Schedule management software: This section states the schedule management software to be used in the program, as it helps to manage and control the schedule throughout the program life cycle. The software tool to be used should be appropriate to the scope and complexity of the program.

Program funding schedule: This section describes the funding schedule for the program's revenue and expenses. Money payable to contractors or for other program expenses, for example, cannot be paid unless it is available to the program. Schedule activities may need to occur at the latest possible start in order to defer expenses until as close as possible to when the funding will be available.

Schedule tracking and monitoring methods: This section describes the methods to use to track and monitor schedule performance according to the master program schedule. Methods to track and monitor all high-level component and program activities and milestones against planned timelines should be stated. If earned value is to be used, it should be stated in this section, including the method to use to report progress.

Schedule metrics: This section describes how schedule progress is to be measured in the program. For example, if the schedule is behind by 10%, does this mean the overall schedule is red, yellow, or green? This section describes what will be measured and how status will be reported against these metrics.

Approvals: This section contains the written approval of the schedule management plan by the program sponsor, program manager, program management office director, members of the Governance Board, and other stakeholders.

Schedule Management Plan Template

<Insert Program Name>
Schedule Management Plan

Program name:	
Program manager:	PM's email address here as a hyperlink
Program sponsor:	
Actual start date:	
Approved end date:	
Program number:	
Revision history:	
Business unit:	

A. PURPOSE

A brief introductory statement defining the purpose of the schedule management plan, such as:

The schedule management plan identifies an agreed-upon sequence of component deliveries (projects and non-project work) to enable them to be planned and managed. It sets forth a scheduling methodology, states the scheduling tool to be used, and sets the format and criteria for monitoring and controlling the program schedule.

B. PROGRAM SCHEDULE CONSTRAINTS

This section describes constraints, both internal and external, that may affect the program schedule.

C. PROGRAM SCHEDULE STANDARDS

This section describes the organization's standards that require consideration as the schedule is developed.

D. DELIVERY OF INCREMENTAL BENEFITS

This section describes the order in which the incremental benefits of the program are to be delivered. It also contains component milestones to show when program benefits are expected to be realized.

E. SCHEDULE MANAGEMENT SOFTWARE

This section states the schedule management software to be used in the program, as it helps to manage and control the schedule throughout the program life cycle.

F. PROGRAM FUNDING SCHEDULE

This section describes the funding schedule for the program's revenue and expenses.

_____

_____

_____

G. SCHEDULE TRACKING AND MONITORING METHODS

This section describes the methods to use to track and monitor schedule performance according to the master program schedule. If earned value is to be used, it is stated in this section along with the method to use to report progress.

_____

_____

_____

H. SCHEDULE METRICS

This section describes how schedule performance is to be measured in the program. It states what will be measured and how status will be reported against these metrics.

_____

_____

_____

I. APPROVALS

This section contains the approval of the schedule management plan by the program sponsor, program manager, program management office director, members of the Governance Board, and other key stakeholders.

SIGNATURES AND DATE APPROVAL OBTAINED

Program manager   _____

Program sponsor   _____

Program management office director   _____

Governance Board chairperson   _____

Governance Board member 1   _____

Governance Board member 2   _____

Governance Board member N   _____

Stakeholder 1   _____

Stakeholder 2   _____

Stakeholder N   _____

Program Financial Plan

The program financial plan helps deal with the myriad of challenges that arise in the financial management arena, some of which can be fatal if not properly managed. It is noted in section 8.2.3, Program Financial Plan Development, in the *Standard for Program Management*—Third Edition (2013).

Program Financial Plan Instructions

The program financial plan will include the following:

Purpose: A brief introductory statement defining the purpose of the financial plan for the program, such as:

> The financial plan documents all the program's financial aspects, as it describes the funding schedules and milestones, baseline budget, contact payments and schedules, financial reporting processes and mechanisms, and financial metrics.

> This plan takes into account items such as risk reserves, potential cash flow problems, international currency rate fluctuations, future interest rate increases or decreases, local laws as appropriate, trends in the cost of materials and supplies, and contract incentives and penalty clauses. It also describes any funds to be retained in contracts as invoices are submitted for payment in those situations in which the contract terms and conditions contain a clause that specifies that a certain percent of the price will be retained and paid once the contract has been completed. It is necessary to recognize that programs may have multiple sources of funding, tend to be long in duration, may have multiple contractors, and may involve international work. Many of the environmental factors that affect this plan may be outside of the program manager's direct control. Further, although the plan is drafted early in the program, it is an iterative document that should be further refined as the program progresses through the life cycle.

> In the Project Management Institute's *Standard for Program Management*—Third Edition (2013), the program financial plan is discussed in section 8.2.3 and builds on the financial framework. It is used as described in section 8.2.4, Component Cost Estimation; 8.2.5, Program Cost Budgeting; and section 8.2.6, Program Financial Monitoring and Control.

Financial framework and goals: This section identifies the overall financial environment for the program, as it pinpoints funds to be expended at key milestones and helps to ensure that money is spent efficiently with the least waste possible.

Funding constraints: This section describes methods to receive program funding since it is rare that 100% of the required funding is provided when the program is initiated. It describes release of funds at specific milestones or on an annual or fiscal year basis. Other examples of constraints are if it is a global program, whether payment is to be made in local currencies; future changes in material prices; or, for contracts, the percent of payment that is to be retained until the contractor's work is complete.

Program operational costs: This section describes the specific costs associated with the operation of the program. It includes infrastructure costs, such as required facilities and systems, or the costs associated with the program management office, including human resources.

Format and criteria for cost estimating: This section describes the various tools and techniques to be used to estimate the costs associated with the program in order that an overall cost estimate for the program can be prepared. Typically, the cost estimate will be prepared in stages given the duration and complexity of the program. Once the cost estimate is approved, it then becomes the basis for the program budget, as the budget baseline is the target for measuring overall financial progress.

Program payment schedule: This section identifies the schedule and milestones where funding for the program is received from the funding organization. This section is needed because program funds typically are spent in advance of receipt of revenues or benefits realization.

Program financial metrics: This section describes the specific financial metrics for the program by which the program's benefits are measured. These metrics are necessary because as changes to schedules and costs occur to the program, these metrics are used against those set at the time the program was approved to help determine whether or not it should be continued. Examples of metrics for consideration are funds expended against total funds available, earned value measurements, the use of contingency and management reserves, return on investment versus that stated in the business case, and the payback period versus that stated in the business case.

Methods for financial monitoring and control: This section describes the approach to be used to identify factors that may create changes to the financial baseline, methods to monitor the environment for changes that may affect the financial plan, ways to monitor the disbursement of funds to contractors, methods to identify impacts to program components from overruns or under-runs, methods to manage the program's infrastructure, and methods to communicate the financial situation to the program stakeholders.

Remaining budget allocations: This section describes the process to return any remaining budget allocations to the overall program budget. It is a consideration during the close program procurement process when all approved payment requests have been processed.

Approvals: This section contains the written approval of the financial plan by the program sponsor, program manager, program management office director, members of the Governance Board, and other stakeholders.

Program Financial Plan Template

<Insert Program Name>
Financial Plan

Program name:	
Program manager:	PM's email address here as a hyperlink
Program sponsor:	
Actual start date:	
Approved end date:	
Program number:	
Revision history:	
Business unit:	

A. PURPOSE

A brief introductory statement defining the purpose of the financial plan, such as:

> The financial plan documents all the program's financial aspects as it describes the funding schedules and milestones, baseline budget, contact payments and schedules, financial reporting processes and mechanisms, and financial metrics.

_____

_____

_____

B. FINANCIAL FRAMEWORK AND GOALS

This section identifies the overall financial environment for the program as it pinpoints funds to be expended at key milestones and helps ensure money is spent efficiently with the least waste possible.

_____

_____

_____

C. FUNDING CONSTRAINTS

This section describes the methods by which program funding will be received since it is rare that 100% of the required funding will be available at the time the program is initiated. Other financial constraints also are included in this section.

_____

_____

_____

D. PROGRAM OPERATIONAL COSTS

This section describes the specific costs associated with the operation of the program, such as infrastructure costs—facilities and systems—and costs associated with the program management office, including human resources.

_____

_____

_____

E. FORMAT AND CRITERIA FOR COST ESTIMATING

This section describes the various tools and techniques to be used to estimate the costs associated with the program in order that an overall cost estimate for the program can be prepared.

_____

_____

_____

F. PROGRAM PAYMENT SCHEDULE

This section identifies the schedule and milestones where funding for the program is received from the funding organization.

_____

_____

_____

G. PROGRAM FINANCIAL METRICS

This section describes the specific financial metrics for the program by which the program's benefits are measured.

_____

_____

_____

H. METHODS FOR FINANCIAL MONITORING AND CONTROL

This section describes the approach to use to identify factors that may create changes to the financial baseline, methods to monitor the environment for changes that may affect the financial plan, ways to monitor the disbursement of funds to contractors, methods to identify impacts to the program components from over-runs and under-runs, methods to manage the program's infrastructure costs, and methods to communicate the financial situation to program stakeholders.

_____

_____

_____

I. REMAINING BUDGET ALLOCATIONS

This section describes the process to return any remaining budget allocations to the overall program budget, which may occur during the close procurement process.

_____

_____

_____

J. APPROVALS

This section contains the approval of the financial plan by the program sponsor, program manager, program management office director, members of the Governance Board, and other key stakeholders.

SIGNATURES AND DATE APPROVAL OBTAINED

Program manager   _____

Program sponsor   _____

Program management office director   _____

Governance Board chairperson   _____

Governance Board member 1   _____

Governance Board member 2   _____

Governance Board member N   _____

Stakeholder 1   _____

Stakeholder 2   _____

Stakeholder N   _____

Procurement Management Plan

The procurement management plan is discussed in section 8.4.1, Program Procurement Planning in the *Standard for Program Management*—Third Edition (2013). It describes the procurement process, including what is to be procured and the means of arriving at a contract. It defers to the contracts management plan, as noted in section 8.4.2, Program Procurement, for the purpose of contracts administration and execution.

Procurement Management Plan Instructions

The procurement management plan includes the following:

Purpose: A brief introductory statement defining the purpose of the procurement management plan, such as:

> The procurement management plan describes the process used to determine what to acquire, such as facilities, goods, materials, and external resources, based on product requirements and describes the contractual methods that will be used.

> This plan should be prepared early in the planning process, as the procurement process is critical to program success and benefits realization. Careful planning and analysis are required and cannot be understated in order that funding is used in the most effective way. Although the plan is drafted early in the program, it is an iterative document that should be further refined as the program moves through the phases of the life cycle, as it may require updates, especially during monitoring and controlling. It also may require updates because of changes in organizational strategy or because of risk responses, which may lead to make-or-buy decisions. This plan also does not replace the procurement management plans to be prepared by the project managers in the program for their specific projects.

> In the Project Management Institute's *Standard for Program Management—Third Edition* (2013), the procurement management plan is described in section 8.4.1 and is in then used as the contracts management plan is developed and may be updated as a result of the contract administration. It uses the PWBS and make-or-buy decision criteria.

Procurement planning approach: This section describes how the program management team will determine which items—facilities, goods, materials, and external resources—should be handled through the procurement process. It describes the make-or-buy process to follow and the various items that should be considered. These include external factors, market conditions, regulations, budget, and available suppliers. It also states the organization's capability to evaluate, procure, and then manage all items that are handled through procurements.

Required and available resources and sourcing decisions: This section describes the specific resources required for the program and whether or not they are available within the organization or need to be procured from external sources.

Required procurements: This section identifies the required procurements for the program. In preparing this section, planning is needed to ensure that there is an emphasis on shared resources across the program so that additional costs are not required later in the program life cycle.

Statements of work for each item to be procured: This section describes statements of work for each item to be procured. It is based on review of the program scope statement and the program work breakdown structure. It is noted that the statements of work are for items at the program level. It may be possible, and desirable, to group items to be procured as one procurement item in a single statement of work. These statements of work then guide the procurement process, as it may be different for the various items that will be procured. They serve to describe in specific detail the procurement item in order that prospective sellers can determine whether or not they can provide the required items. Information in the statement of work may include specifications, quantities required, quality levels, performance requirements, the period of performance, and other requirements as appropriate. This section also states how work will be managed for each item that is procured and whether the work will be done internally or externally.

Procurement documents: This section states the specific procurement documents that will be used for each item to be procured, such as request for proposal, request for quotation, or invitation to bid. The purpose is to structure these documents in a way that prospective sellers can prepare an accurate and complete response, and the program management team can then evaluate the responses easily. The documents should be ones that enable the prospective supplier to propose an alternative means of meeting the requirement if possible to provide greater flexibility.

Contract types: This section defines and documents the specific contract types that will be used. They will vary based on the nature of the procurement. While fixed-price, cost-plus, and time and materials contracts are typical, other vehicles such as basic ordering agreements, other umbrella-type contracts, leasing, and integrated volume discounts should be considered as appropriate.

Qualified seller lists: In order to expedite the procurement process, qualified seller lists may be appropriate. This section describes how these lists will be developed—by others in the organization, such as members of the procurement department or by the program management team. It also states how information about qualified sellers will be obtained. A variety of tools and techniques can be used, such as requests for information, the Internet, advertising, directories, associations, and catalogs. This section also states whether on-site visits will be made as part of the process to develop these lists. It may be useful as well for the program management team to contact seller references to obtain information on past performance.

Proposals: This section describes the processes, procedures, and evaluation criteria that will be used to review proposals from potential sellers. It states who will be involved in evaluating the proposals that are received so that there is no bias in the process. The evaluation criteria may be limited to price for certain items or may be more complex for other items. If the latter, it is appropriate to consider whether prospective sellers understand the requirements, the management approach to be followed, the overall cost, the technical approach to be used, the people who will support the procurement, the schedule for deliverable items, the risk involved, past performance and references, intellectual property and proprietary rights, financial considerations, and warranties.

Approvals: This section contains the written approval of the procurement management plan by the program sponsor, program manager, program management office director, members of the Governance Board, and other stakeholders.

Procurement Management Plan Template

<Insert Program Name>
Program Procurement Management Plan

Program name:	
Program manager:	PM's email address here as a hyperlink
Program sponsor:	
Actual start date:	
Approved end date:	
Program number:	
Revision history:	
Business unit:	

A. PURPOSE

A brief introductory statement defining the purpose of the procurement management plan, such as:

The procurement management plan describes the process used to determine what to acquire, such as facilities, goods, materials, and external resources, based on product requirements and describes the contractual methods that will be used.

B. PROCUREMENT PLANNING APPROACH

This section describes how the program management team will determine which items are to be handled as procurements. It describes the make-or-buy process to follow and the various items that should be considered. It also states the organization's capability to evaluate, procure, and then manage all the items that are procured externally.

C. REQUIRED AND AVAILABLE RESOURCES AND SOURCING DECISIONS

This section describes the specific resources required for the program and whether or not they are available within the organization or need to be procured from external sources.

D. REQUIRED PROCUREMENTS

This section identifies the required procurements for the program.

E. STATEMENTS OF WORK FOR EACH ITEM TO BE PROCURED

This section describes statements of work for each item to be procured. These statements of work guide the procurement process and describe in detail the specific procurement item. This section also states how work will be managed for each item that is procured and whether the work will be done internally or externally.

F. PROCUREMENT DOCUMENTS

This section states the specific procurement documents that will be used for each item to be procured.

_____

_____

_____

G. CONTRACT TYPES

This section defines and documents the specific contract types that will be used based on the nature of the procurement item.

_____

_____

_____

H. QUALIFIED SELLER LISTS

This section describes whether qualified seller lists will be used and how they will be developed. It describes the process that will be followed to obtain information on each of the possible sellers to be on the list.

_____

_____

_____

I. PROPOSALS

This section describes the processes, procedures, and evaluation criteria that will be used to review proposals from potential sellers. It states who will be involved in the evaluation process so there is no bias in the process.

_____

_____

_____

J. APPROVALS

This section contains the approval of the procurement management plan by the program sponsor, program manager, program management office director, members of the Governance Board, and other key stakeholders.

SIGNATURES AND DATE APPROVAL OBTAINED

Program manager _____

Program sponsor _____

Program management office director _____

Governance Board chairperson   _____

Governance Board member 1   _____

Governance Board member 2   _____

Governance Board member N   _____

Stakeholder 1   _____

Stakeholder 2   _____

Stakeholder N   _____

Contracts Management Plan

The contracts management plan builds on the procurement management plan, elaborating on contracts administration to ensure appropriate cost, quality, and schedule. It also is noted in section 8.4.2, Program Procurement in the *Standard for Program Management*—Third Edition (2013).

Contracts Management Plan Instructions

The contracts management plan will include the following:

Purpose: A brief introductory statement defining the purpose of the contracts management plan, such as:

> The contracts management plan describes the process that will be followed to administer each of the program-level contracts to ensure that deliverables meet requirements concerning cost, schedule, and quality.

> This plan builds on the procurement management plan, which is developed earlier in the program. In fact, once the contracts management plan is approved, it can become an appendix to the procurement management plan. It is an iterative document and should be reviewed periodically by the program management team, especially as contracts are closed, and new contracts are awarded to see if changes are warranted. This plan covers the contract administration activities throughout the life of the contract.

> In the Project Management Institute's *Standard for Program Management*—Third Edition (2013), the contracts management plan is noted in section 8.4.2. Updates to this plan may result from as the contracts are under way.

Roles and responsibilities: This section describes the roles and responsibilities of the program management team, the component managers, the procurement department, the legal department, and others in the organization concerning contract administration. It states the actions the program management team can take on its own without involvement from the procurement or legal departments. It also describes how the program management team will interface with the people on various projects in the program responsible for contract administration. Experts may also be needed, especially for global programs in which specific knowledge about contract laws in other countries will apply where the procurement will be awarded. Actions to be taken by whom if conditions outlined in the contract are not met are stated in this section.

Buyer-seller performance requirements: For overall program success, a partnership is needed between the buyer and the sellers. This section states the specific performance requirements each party should meet. For example, it describes the process to review each deliverable. While the deliverable due date is specified in the contract, the buyer should review it according to a predetermined schedule. The seller then should have a certain schedule in order to respond to needed changes if rework is required. Nontechnical factors, such as establishment of trust and excellent working relationships between the parties, also are critical to program success. The buyer enables the seller to complete the contract, pays the agreed-upon price, formally accepts the deliverable, and closes the contract. The seller completes the deliverable according to the contractual terms and conditions.

Monitoring and control: This section describes specific monitoring and controlling processes to be followed. It identifies the circumstances to consider if the contractual agreement is violated, to ensure that the deliverables meet requirements, to determine that the relationship between the buyer and seller ends as stipulated, and to ensure that any post-program communications occur as specified.

Subcontract selection process criteria: Since many prime contractors in a program issue subcontracts, this section states the selection criteria for these subcontracts to ensure consistency in meeting overall program objectives and benefit expectations. If the subcontract involves an initiative in another country, consideration of local laws is necessary. Contractual terms and conditions should be specified for the subcontractors as appropriate.

Schedules: This section describes the process the program management team will use to coordinate the various schedules prepared by the program contractors and to ensure that these schedules are part of the program's master schedule. This section is required because the lack of delivery or performance of one contractor may affect the overall performance of the program and the delivery of the expected program benefits. It states how often revised schedules will be provided by the contractors, and the process to learn of any problems so corrective action can be taken as quickly as possible. It describes the tools and techniques to be used, as it may be beneficial for all the program contractors to use the same scheduling software for technical compatibility.

Meetings to be held with contractors: In many large programs, much of the work may be done by contractors. If this is the case in the program, this section would describe whether any meetings are planned to be held with the various contractors to better coordinate the work to be done and to resolve any conflicts that could occur during the process. The program management team then can use these meetings to coordinate and resolve any scheduling or other constraints.

Status reporting: This section describes the status reports to be provided by the various program contractors to describe progress in terms of achieving contractual objectives, and how the individual reports then will be merged into an overall program status report. It states the frequency in which the reports will be prepared and the format to be used. Ideally, each contractor should follow a standard format, which should include as a minimum schedule, budget, scope, risk, benefits, and quality metrics and approval status of program deliverables.

Contract performance review meetings: The section describes the items to be covered in contractor performance reviews and how often these meetings will be held. These meetings are useful not only to assess performance to date but also to determine if product pricing has changed significantly and requires revisions.

Change control: Change requests may be needed because of numerous factors, such as a change in overall strategy, a change in requirements, performance issues, etc. This section describes the process to follow when a change is needed to a contract in the program. It describes the contract change control system, as it states when change requests are required, the format to use to submit a change request, who must analyze requests for their impact on other aspects of the program, documentation requirements, and who is authorized to approve change requests. It also describes the process to use to notify the contractors when a change request has been approved or rejected.

Payment control: This section describes the payment control system that will be used to ensure that all invoices are processed in a timely fashion, that contract terms are followed, and that duplicate payments are not made. It describes the necessary approvals before a payment can be made.

Contract closure: This section describes the specific process the program team will use to close contracts and to determine that all contractual obligations have been met. It describes the specific documents to be reviewed, documents that will be archived, how lessons learned will be conducted, any performance reviews that may be held, contract closure audits that may be conducted, stakeholders to be notified, and any required follow-up work or warranties. This section also describes the process to follow if contracts are terminated early, such as documenting why the contract was terminated, determining the extent of the actual work that was performed, and updating contractual records.

Approvals: This section contains the written approval of the contracts management plan by the program sponsor, program manager, program management office director, members of the Governance Board, and other stakeholders.

Contracts Management Plan Template

<Insert Program Name>
Contracts Management Plan

Program name:	
Program manager:	PM's email address here as a hyperlink
Program sponsor:	
Actual start date:	
Approved end date:	
Program number:	
Revision history:	
Business unit:	

A. PURPOSE

A brief introductory statement defining the purpose of the contracts management plan, such as:

> The contracts management plan describes the process that will be followed to administer each of the program-level contracts to ensure that deliverables meet requirements concerning cost, schedule, benefits, and quality.

_____

_____

_____

B. ROLES AND RESPONSIBILITIES

This section describes the roles and responsibilities of the program management team, the component managers, the procurement department, the legal department, and others in the organization concerning contract administration. It states the actions the program management team can take on its own without involvement from other groups and describes how the program management team will interface with the various people administering contracts at the project level. It describes whether the use of experts will be required.

_____

_____

_____

C. BUYER-SELLER PERFORMANCE REQUIREMENTS

This section states the specific performance requirements of the buyer and the seller. It describes the process to follow in reviewing deliverables and the need for attention to nontechnical factors.

D. MONITORING AND CONTROL

This section describes specific monitoring and controlling processes to be followed. It identifies the circumstances to consider if agreements are violated, to ensure deliverables meet requirements, to see that the relationship between the buyer and seller ends as stipulated, and to see if post-program communications occur as specified.

E. SUBCONTRACT SELECTION PROCESS CRITERIA

This section describes the selection criteria for any subcontracts by prime contractors to ensure consistency in meeting overall program objectives and benefit expectations. It specifies contractual terms and conditions to be included in the subcontracts as appropriate.

F. SCHEDULES

This section describes the process the program management team will use to coordinate the various schedules prepared by the program contractors and to ensure that these schedules are part of the program's master schedule. It states how often revised schedules will be provided by the contractors, and the process to learn of any problems so corrective action can be taken. It describes tools and techniques to be used for technical compatibility.

G. MEETINGS TO BE HELD WITH CONTRACTORS

This section describes whether any meetings are planned to be held with the various contractors to better coordinate the work to be done and to resolve any conflicts that could occur in the process.

_____

_____

_____

H. STATUS REPORTING

This section describes status reports to be provided by the program contractors to describe progress in terms of achieving contractual objectives and how the individual reports will be merged into an overall program status report. It states the reporting frequency and the format to be used for the reports.

_____

_____

_____

I. CONTRACT PERFORMANCE REVIEW MEETINGS

This section describes how often contractor performance will be reviewed by the program team and the items to be covered in a typical review.

_____

_____

_____

J. CHANGE CONTROL

This section describes the contract change control system to be followed in the program, including notification to the contractors when a change request has been approved or rejected.

_____

_____

_____

K. PAYMENT CONTROL

This section describes the payment control system to be used to ensure that invoices are processed in a timely way, that contract terms and conditions are followed, and that duplicate payments are not made. It describes the needed approvals before a payment is made.

_____

_____

_____

_____

L. CONTRACT CLOSURE

This section describes the process the program team will use to close contracts and to determine that all contractual obligations have been met. It also describes the process to follow if contracts are terminated early.

_____

_____

_____

M. APPROVALS

This section contains the approval of the contracts management plan by the program sponsor, program manager, program management office director, members of the Governance Board, and other key stakeholders.

SIGNATURES AND DATE APPROVAL OBTAINED

Program manager _____

Program sponsor _____

Program management office director _____

Governance Board chairperson _____

Governance Board member 1 _____

Governance Board member 2 _____

Governance Board member N _____

Stakeholder 1 _____

Stakeholder 2 _____

Stakeholder N _____

Interface Management Plan

The interface management plan is not in either the Second or Third Editions of the *Standard for Program Management*. However, throughout the Third Edition (2013), the need for the program manager to focus on the interfaces between the program components and also the interfaces at the portfolio level to ensure the program remains in alignment with organizational strategies is emphasized. We feel it is useful for those programs that have complex interfaces and as a stimulus for consideration of the matters it addresses on smaller programs.

If multiple individuals are taking care of relationships with a given external organization on the same matters, conflicts could result. Neglected interfaces with known stakeholders are almost always problematic. The interface management plan addresses these issues and more.

Interface Management Plan Instructions

The interface management plan includes the following:

Purpose: A brief introductory statement defining the purpose of the interface management plan, such as:

> The interface management plan identifies and maps interrelationships within the program, with other programs in the portfolio, and with factors outside of the program. It identifies interfaces in terms of ones that are organizational, technical, interpersonal, logistical, and political.

It addresses internal and external interfaces as well as interdependencies and interrelationships. It then assists in identifying risks with these interrelationships throughout the management of the program. It describes roles and responsibilities for program interface management. Although the plan is drafted early in the program, it is an iterative document that should be further refined as the program moves through the life cycle.

Identification of organizational interfaces: This section defines each organizational interface in the program, such as the various departments or units within the organization to be involved in the program, the current relationships among them, and the informal and formal relationships among them. It also describes the relationship with the portfolio manager and the program management office.

Identification of technical interfaces: This section states the different disciplines and specialties that are needed to complete the program. It includes items such as software, subsystems, testing, and equipment that may need to be coordinated. It also should describe whether there are any transitions within the program management life cycle that may present challenges.

Identification of interpersonal interfaces: This section describes the formal and informal reporting relationships among members of the program team. It includes any cultural or language differences that may affect these working relationships and describes whether any levels of trust and respect already exist among team members. It also states customer-supplier relationships.

Identification of logistical interfaces: This section describes logistical interfaces, such as the use of virtual teams in the program and whether stakeholders are in different locations, time zones, or countries.

Identification of political interfaces: This section describes the individual goals and objectives of program stakeholders, to especially note whether people have informal power in areas of importance to the program. It lists any informal alliances that may exist.

Management approach for component interfaces: This section describes how interfaces between the various program components will be managed. Changes often can occur because of interfaces with various program components (projects and non-project work), and a change to one component may affect another component in the program.

Management approach for interfaces with the program's Governance Board: This section describes the approach to follow to manage interfaces with other enterprises, such as customers, suppliers, and other parts of the organization, which are typically handled through work with the program's Governance Board.

Management approach for interfaces with other aspects of the program: This section describes how interface planning will be coordinated with other aspects of the program, such as with schedule development, human resource planning, communications planning, the program work breakdown structure, and risk management planning and analysis.

Roles and responsibilities for interface management: This section states the various roles and responsibilities for interface management in the program.

Approvals: This section contains the written approval of the interface management plan by the program sponsor, program manager, program management office director, members of the Governance Board, and other stakeholders.

Interface Management Plan Template

<Insert Program Name>
Program Interface Management Plan

Program name:	
Program manager:	PM's email address here as a hyperlink
Program sponsor:	
Actual start date:	
Approved end date:	
Program number:	
Revision history:	
Business unit:	

A. PURPOSE

A brief introductory statement defining the purpose of the interface management plan, such as:

> The interface management plan identifies and maps interrelationships within the program, with other programs in the portfolio, and with factors outside of the program. It identifies interfaces in terms of ones that are organizational, technical, interpersonal, logistical, and political.

_____

_____

_____

B. IDENTIFICATION OF ORGANIZATIONAL INTERFACES

This section defines each organizational interface in the program, such as the various departments or units within the organization to be involved in the program, the current relationships among them, and the informal and formal relationships among them.

_____

_____

C. IDENTIFICATION OF TECHNICAL INTERFACES

This section states the different disciplines and specialties that are needed to complete the program. It includes items such as software, subsystems, testing, and equipment that may need to be coordinated. It also should describe whether there are any transitions within the program management life cycle that may present challenges.

_____

_____

_____

D. IDENTIFICATION OF INTERPERSONAL INTERFACES

This section describes the formal and informal reporting relationships among members of the program team, including any cultural or language differences that may affect these working relationships and any levels of trust and respect that already exist among team members. It also states customer-supplier relationships.

_____

_____

_____

E. IDENTIFICATION OF LOGISTICAL INTERFACES

This section describes logistical interfaces, such as the use of virtual teams in the program and whether stakeholders are in different locations, time zones, or countries.

F. IDENTIFICATION OF POLITICAL INTERFACES

This section describes the individual goals and objectives of program stakeholders, to especially note whether people have informal power in areas of importance to the program. It lists any informal alliances that may exist.

G. MANAGEMENT APPROACH FOR COMPONENT INTERFACES

This section describes how interfaces between the various program components (projects and non-project work) will be managed.

H. MANAGEMENT APPROACH FOR INTERFACES WITH THE PROGRAM GOVERNANCE BOARD

This section describes the approach to follow to manage interfaces with other enterprises, such as customers, suppliers, and other parts of the organization, which are typically handled through work with the program's Governance Board.

I. MANAGEMENT APPROACH FOR INTERFACES WITH OTHER ASPECTS OF THE PROGRAM

This section describes how interface planning will be coordinated with other aspects of the program, such as with schedule development, human resource planning, communications planning, the program work breakdown structure, and risk management planning and analysis.

_____

_____

_____

J. ROLES AND RESPONSIBILITIES FOR INTERFACE MANAGEMENT

This section states the various roles and responsibilities for interface management in the program.

_____

_____

K. APPROVALS

This section contains the approval of the interface management plan by the program sponsor, program manager, program management office, members of the Governance Board, and other key stakeholders.

Signatures and Date Approval Obtained

Program manager _____

Program sponsor _____

Program management office director _____

Governance Board chairperson _____

Governance Board member 1 _____

Governance Board member 2 _____

Governance Board member N _____

Stakeholder 1 _____

Stakeholder 2 _____

Stakeholder N _____

Program Management Plan

According to the *Standard for Program Management*—Third Edition (2013), the program management plan is described in section 7.1.1.2, Program Preparation, and it includes candidate program components and the management plans that are needed to achieve the program's objectives. It does not list all of these management plans *per se*, and each organization can then decide which plans should be included considering the plans described in this chapter.

The subsidiary plans the *Standard for Program Management*—Third Edition (2013) does mention in section 8.3.2, Program Management Plan Development are the:

- Program financial management plan
- Benefits realization plan
- Stakeholder engagement plan
- Governance plan
- Communications management plan
- Scope management plan
- Schedule management plan
- Resource management plan
- Procurement management plan
- Program documentation archive plan [discussed as part of the knowledge management plan in governance chapter]

Integrating these subsidiary plans provides the basis to set the overall framework for managerial controls and determines how best to manage and control the program's components. It also notes that updates to the program management plan may be required:

- When baseline are established
- When certain types of change requests are approved

The *Standard for Program Management*—Third Edition describes that this plan is developed based on the organization's strategic plan, the business case, and the program charter. It also mentions "other outputs" (section 7.1.1.2) but does not specify them. We would suggest the: program roadmap, scope statement, program work breakdown structure, schedule, and metrics. Once the program plan is approved, since it is the primary outcome of the program definition phase, the program then moves according to the Standard to the program benefit delivery phase. The program management plan then is a reference document that is used to measure overall program success. Its objective is to ensure the program remains aligned with the organization's strategy in order to deliver its proposed benefits.

The items in the program management plan may be separate documents or some or all of them could be placed in a single volume, if desired, for smaller programs. In some cases, one of these plans could simply point to a standard organization plan. Regardless, the program management plan establishes the approach the program team will follow to contribute toward achievement of the organization's strategic goals and objectives.

Program Management Plan Instructions

The program management plan includes the following:

Purpose: A brief introductory statement defining the purpose of the program management plan, such as:

The program management plan provides direction as to how the program will be managed, monitored and controlled, and closed; guides the allocation and use of program resources; and defines the program's deliverables and benefits.

Once the program management plan is approved, program execution or delivery of the program's benefits begins. This plan contains the various subsidiary plans, and they can be set up as attachments to the program management plan, or they can be consolidated into a single document. Although this plan is prepared early in the program's life cycle, it is an iterative document that is expected to change throughout the program as the program progresses. It also is a starting point for project managers working on the program as they develop the plans for their specific projects, as it provides overall guidance. If the program begins after several components have been under way, the existing project plans should be reviewed as this plan is developed.

The *Standard for Program Management*—Third Edition (2013) suggests that some long programs may lead to an increase in opportunities throughout the program's duration, and in these cases, it may be appropriate to state an innovation strategy as a component to the plan. For example, at the time the program was initiated, certain components were approved to be part of the program as shown in the roadmap. But, as the program continues, other projects may need to be added or replaced with ones that would lead to more benefits. This innovation strategy helps to compensate for the changing environment in which the program will be executed and may require a separate budget to compensate for any possible threats to the program or even a decrease in efficiency.

Program end result and schedule: This section describes the key program deliverables and their schedule for completion. It also describes the major program benefits. It provides background information, including why the program was initiated, and how it aligns with the organization's mission, vision, and values.

Program budget and cash flow requirements: This section contains information on the budget that is needed in order to complete the program. It describes cash flow requirements, referring to the financial management plan for details.

Program risks and issues: This section presents an overview of the key program risks, opportunities, and issues. It notes that details are included in the program risk management plan and risk register.

Program dependencies: This section describes the key dependencies between this program and other programs and projects under way by the organization.

Program assumptions: This section lists the assumptions or factors that the program management team considers to be true, real, or certain that are part of this plan. Subsidiary plans will describe some assumptions in greater detail as appropriate.

Program constraints: This section describes constraints, or restrictions or limits, that may affect overall program performance. These constraints are described in greater detail in some of the subsidiary program plans as appropriate.

Program management approach: This section describes the approach the program management team plans to follow as it executes the program and the methodology, tools, and techniques to be used. It describes how performance will be measured. If the organization has an existing program management methodology, it states whether or not it will be used as is or whether it requires tailoring. It also describes how the program management team will work with component managers on the program, functional managers in the organization, the program's Governance Board, other stakeholders, and customers. It describes the metrics that will be used to help track and monitor the program to ensure deliverables are completed as planned, and expected benefits are realized. It shows how the various plans are integrated and coordinated at the program level. It also states how the program management team will document and use lessons learned throughout the program. Any supporting feasibility studies (technical, economic, ethical, social, environmental, etc.) should be attached to this section.

Program organization: This section states how the program will be organized. It describes the composition of the program management core team, the use of a program management office, and the interaction with the program's Governance Board.

Program management information system: This section describes the approach the program management team will use to collect and manage all program-related information. It shows the various tools and processes that will be used, such as software, documentation management, configuration management, knowledge repositories, earned value management, risk data base and analysis, financial management systems, requirements management activities and tools, and the use of a program management office. It describes roles and responsibilities for use of this system and access privileges to it. This system is used for overall program performance monitoring and control, so this section also describes the specific reports that will be prepared and how they will be distributed, following the program communications management plan.

Schedule milestones and budget goals for program components: This section describes the high-level schedule milestones for each of the program's components (projects and non-project work) and also includes the allocated budget for each component.

Tolerance ranges and decision making: Each organization has different levels of tolerance for variances in cost, scope, schedule, and risks and in how they will be handled. This section describes the level of tolerances for the program based on expectations of stakeholders and members of the program's

Governance Board. It also states approvals required if the tolerance levels are exceeded. It discusses the various types of program decisions that are expected to be made by the program manager based on items in the program charter and the program manager's level of authority and accountability, and those that require escalation to the Governance Board. It states how the results of the decisions will be presented, recorded, communicated and managed, using the program communications management plan as a framework.

Updates to the program management plan: This section states when updates to the program management plan are required, such as when there are changes in the overall strategic direction of the organization, in the program's requirements, in organizational processes and procedures, or to the program's scope, schedule, cost, or quality baselines; unplanned circumstances; or risks that turn into issues that then require updates to the plan. It describes the documentation management process to follow to update the plan and how key stakeholders will be notified that the plan has been updated. A formal change request should be prepared and approved before the plan is updated.

Approvals: This section contains the written approval of the program management plan by the program sponsor, program manager, program management office director, members of the Governance Board, and other stakeholders.

Program Management Plan Template

<Insert Program Name>
Program Management Plan

Program name:	
Program manager:	PM's email address here as a hyperlink
Program sponsor:	
Actual start date:	
Approved end date:	
Program number:	
Revision history:	
Business unit:	

A. PURPOSE

A brief introductory statement defining the purpose of the program management plan, such as:

The program management plan provides direction as to how the program will be managed, monitored and controlled, and closed; guides the allocation and use of program resources; and defines the program's deliverables and benefits.

_____

_____

_____

B. PROGRAM END RESULT AND SCHEDULE

This section describes the key program deliverables and their schedule for completion. It also describes key program benefits and presents background information as to why the program was initiated.

_____

_____

_____

C. PROGRAM BUDGET AND CASH FLOW REQUIREMENTS

This section contains information on the budget that is needed to complete the program and describes cash flow requirements, referring to the financial management plan for details.

_____

_____

_____

D. PROGRAM RISKS AND ISSUES

This section presents an overview of the key program risks, opportunities, and issues. It notes that details are contained in the program risk management plan and risk register.

_____

_____

_____

E. PROGRAM DEPENDENCIES

This section describes the key dependencies between this program and other programs and projects under way in the organization.

_____

_____

_____

F. PROGRAM ASSUMPTIONS

This section lists the assumptions or factors that the project management team considers to be true, real, or certain that are part of this plan.

_____

_____

_____

G. PROGRAM CONSTRAINTS

This section lists the constraints, or restrictions or limits, that may affect program performance.

_____

_____

_____

H. PROGRAM MANAGEMENT APPROACH

This section describes the approach the program management team plans to follow as it executes the program, and the methodology, tools, and techniques to be used. It shows how the various plans are integrated and coordinated at the program level. Supporting feasibility studies, if applicable, are attached to this section.

_____

_____

_____

I. PROGRAM ORGANIZATION

This section states how the program will be organized. It describes the composition of the program management core team, the involvement of the program management office, and the interaction with the program's Governance Board.

_____

_____

_____

J. PROGRAM MANAGEMENT INFORMATION SYSTEM

This section describes the approach the program management team will use to collect and manage all program-related information. It describes the various tools and processes that will be used and the roles and responsibilities for use of the system and access privileges to it. It states the performance reports to be prepared and how they will be distributed, following the program communications management plan.

K. SCHEDULE MILESTONES AND BUDGET GOALS FOR PROGRAM COMPONENTS

This section describes the high-level schedule milestones for each of the program's components and includes the allocated budget for each component.

L. TOLERANCE RANGES AND DECISION MAKING

This section describes the tolerance levels for the program based on stakeholder expectations and those of the members of the Governance Board in terms of variances in cost, scope, schedule, and risk, and how the variances will be handled. It also states approval levels if the tolerance levels are exceeded. It describes the various types of program decisions the program manager is expected to make without the need to escalate them to the Governance Board. It describes how the results of these decisions will be communicated and managed, using the program communications management plan as a framework.

M. UPDATES TO THE PROGRAM MANAGEMENT PLAN

This section states when updates to the program management plan are required, the documentation process to follow to update the plan, and how key stakeholders will be notified that the plan has been updated.

N. APPROVALS

This section contains the approval of the program management plan by the program sponsor, program manager, program management office director, members of the Governance Board, and other key stakeholders.

SIGNATURES AND DATE APPROVAL OBTAINED

Program manager _____

Program sponsor _____

Program management office director _____

Governance Board chairperson _____

Governance Board member 1 _____

Governance Board member 2 _____

Governance Board member N _____

Stakeholder 1 _____

Stakeholder 2 _____

Stakeholder N _____

Section 7C: Program Benefits Delivery: Executing the Program

> Good fortune is when opportunity meets with planning.
>
> **—Thomas Alva Edison**

Having said that plans are useless, but also that planning is tremendously important, now we talk about them coming together. But then we remember that as living documents, our plans and processes are designed to systematically evolve into desired reality.

It is useful to consider the program benefits delivery phase, albeit just one phase, in two pieces—the executing part and the monitoring and controlling part. The bulk of program work is in the execution, the subject of this chapter. In the *Standard for Program Management*—Third Edition (2013), this phase consists of three sub-phases:

- Component planning and authorization (discussed in this book in Chapter 6 on Governance)
- Component oversight and integration (discussed in Chapter 6 and in Section 7D on Monitoring and Controlling the Program)
- Component transition and closure (discussed in Chapter 6 and in Section 7E on Closing the Program)

In PMI's *Examination Content Outline* (PMI, 2011), a high-level view of the tasks involved in Executing the Program follows:

- Charter and initiate projects
- Establish consistency with uniform standards, resources, tools, and processes for informed decisions
- Establish a communications feedback mechanism to capture lessons learned during the program
- Lead human resources to improve team engagement and achieve commitment to the program's goals
- Review the performance of project managers in executing their projects
- Execute the appropriate program management plans
- Consolidate project and program data to monitor and control performance and communicate to stakeholders
- Evaluate the program's status
- Approve closure of projects once deliverables are complete

Team Charter

A successful and motivated team is essential to success. A team can benefit greatly from a team charter, although it is not included in the *Standard for Program Management*—Third Edition (2013). This is true as it provides an agreed-upon model and standards for intra-team interactions not covered by other plans.

Team Charter Instructions

The team charter formalizes the program team members' roles and responsibilities. It provides guidelines for the team's operations. It describes the processes that the team will use to complete the program. It states the expectations to be met by the team members, such as the specific responsibilities regarding the program schedule and the realization of program benefits. It includes procedures to resolve conflicts and describes how the team will plan the work, participate in making decisions, and communicate throughout the program. The team charter includes the following:

Purpose: A brief introductory statement defining the purpose of the team charter, such as:

> The team charter states the team members' roles and responsibilities and provides guidelines for the team's operation throughout the life of the program. It describes actions and decisions the team can take on its own and escalation processes to follow as required.

> The team charter formalizes the program team members' roles and responsibilities. It provides guidelines for the team's operations. It describes the processes that the team will use to complete the program. It states the expectations to be met by the team members, such as the specific expectations regarding the program schedule and benefits realization. It includes procedures to resolve conflicts and describes how the team will plan the work, participate in making decisions, and communicate throughout the program.

Program commitment statement: This section describes the team's commitment to the program in terms of the team's desire to achieve the program's objectives and to be able to deliver the expected benefits of the program.

Program manager's role: The charter lists the program manager's name and describes how the team will interact with the program manager during the program. It contains specific roles and responsibilities of the various team members in terms of interaction with the program manager, recognizing that some team members will have more interaction than others.

Program sponsor's role: The charter lists the program sponsor's name and describes how the team will interact with the sponsor during the program.

Client's role: This section names the client and states how the program management team will interact with the client during the program.

Other stakeholders involved in the program: As described in the stakeholder engagement plan, a variety of stakeholders will be involved in the program at different times during the program's life cycle. This section describes the level of interaction of the program team members with the stakeholders during the program. A responsibility assignment matrix may be attached to this section.

Team performance objectives: This section describes specific performance objectives of the program team members during the program's life cycle. It links the performance objectives to the program's objectives.

Success measures: While there are overall success measures in terms of on time, within budget, and within specification delivery of the program's product, service, and result as well as delivery of the program's benefits, this section describes specific success measures in terms of overall team performance.

Scope/boundaries of the team's work: This section builds on the scope management plan and the scope statement to describe any boundaries that may affect the work of the program team. It lists specific constraints and assumptions to be considered.

Deliverables: This section uses the program work breakdown structure and lists the various program packages. A responsibility assignment matrix may be attached to this section to show the roles and responsibilities of the program team members for the various deliverables.

Conflict management process: Recognizing that programs inevitably will have conflicts, this section describes a process to resolve conflicts that may exist among the team members. It notes how the team can resolve conflicts without the need to involve the program manager and states the process to follow if involvement from the program manager is required. It also describes how team members will learn of the conflict resolution. For example, if the conflict only involves a few team members, this section states whether the entire team should be included in the resolution method that is used.

Decision-making process: This section describes the process the team will follow to make program decisions. It states the types of decisions the team can make on its own without the need to involve the program manager and how these decisions will be communicated throughout the program and to relevant stakeholders according to the communications management plan. It also states when the program manager needs to be consulted before a decision is made and the circumstances when the team needs to escalate a decision to the program manager.

Administrative activities: Each program has a variety of administrative activities that must be performed for overall program success. This section states the roles and responsibilities of the team members in terms of these administrative activities. For example, it describes responsibilities in areas such as preparing agendas for various meetings, taking minutes, distributing action items, maintaining various logs and registers, updating plans and other documents, updating the knowledge management repository, and performing closeout tasks. A responsibility assignment matrix for these administrative activities may be attached to this section.

Issue escalation process: Issues are common in programs. In addition to use of the issue register to identify, track, monitor, and resolve issues, this section of the charter states the process the team will use to resolve issues on its own. It also states when issues should be raised to the program manager for consultation or for actual resolution.

Updates to the charter: This section states how often the charter will be reviewed to determine whether updates to it are required and who will be responsible for these updates. For example, if the conflict resolution process is not being followed or is not effective, the charter will require updates. As new members join the team, they may have ideas for changes to the charter, and as members leave the team, they should comment as to whether or not they feel changes to the charter are warranted.

Approvals: This section lists the approval of the team charter by the program sponsor, program manager, and each team member. When new team members join the team, they should sign the charter to indicate their commitment to it.

Team Charter Template

<Insert Program Name>
Team Charter

Program name:	
Program manager:	PM's email address here as a hyperlink
Program sponsor:	
Actual start date:	
Approved end date:	
Program number:	
Revision history:	
Business unit:	

A. PURPOSE

A brief introductory statement defining the purpose of the team charter, such as:

The team charter states the team members' roles and responsibilities and provides guidelines for the team's operation throughout the life of the program. It describes actions and decisions the team can take on its own and escalation processes to follow as required.

_____

_____

_____

B. PROGRAM COMMITMENT STATEMENT

This section describes the team's commitment to the program in terms of the team's desire to achieve program objectives and deliver the program's expected benefits.

_____

_____

_____

C. PROGRAM MANAGER'S ROLE

This section lists the program manager's name and describes how the team will interact with the program manager during the program.

_____

_____

_____

D. PROGRAM SPONSOR'S ROLE

This section states the program sponsor's role and describes how the team will interact with the sponsor during the program.

_____

_____

_____

E. CLIENT'S ROLE

This section names the client and states how the team will interact with the client during the program.

_____

_____

_____

F. OTHER STAKEHOLDERS INVOLVED IN THE PROGRAM

This section describes the level of interaction of the program team members with the stakeholders during the program. A responsibility assignment matrix may be attached to this section.

_____

_____

_____

G. TEAM PERFORMANCE OBJECTIVES

This section describes specific performance objectives of the team members during the program's life cycle to link these objectives to the overall program objectives.

_____

_____

_____

H. SUCCESS MEASURES

This section describes specific success measures in terms of overall team performance.

_____

_____

_____

I. SCOPE/BOUNDARIES OF THE TEAM'S WORK

This section describes any boundaries that may affect the work of the program team and lists specific constraints and assumptions to be considered.

_____

_____

_____

J. DELIVERABLES

This section lists the various program packages in the program's work breakdown structure and shows the roles and responsibilities of the program team members to complete these deliverables. A responsibility assignment matrix may be attached.

_____

_____

_____

K. CONFLICT MANAGEMENT PROCESS

This section describes a process for use among the team to resolve conflicts without the need to involve the program manager and when involvement from the program manager is required. It also describes communication approaches to inform others of how conflicts were resolved.

_____

_____

_____

L. DECISION-MAKING PROCESS

This section describes the process the team will follow to make program decisions and states the types of decisions the team can make on its own without involvement from the program manager and when the program manager should be consulted or should make the decision. It describes how these decisions will be communicated following the communications management plan.

_____

_____

_____

M. ADMINISTRATIVE ACTIVITIES

This section states the roles and responsibilities of team members in terms of the program's administrative activities. A responsibility assignment matrix may be attached to this section.

_____

_____

_____

N. ISSUE ESCALATION PROCESS

This section states the process the team will use to resolve issues and when issues should be raised to the program manager for consultation or resolution.

_____

_____

_____

O. UPDATES TO THE CHARTER

This section states how often the charter will be reviewed to assess its effectiveness and the process the team will follow to update it as needed.

_____

_____

_____

P. APPROVALS

This section contains the approval of the team charter by the program sponsor, program manager, and each team member.

SIGNATURES AND DATE APPROVAL OBTAINED

Program manager   _____

Program sponsor   _____

Team member 1   _____

Team member 2   _____

Team member N   _____

Lessons Learned Review Agenda

Agendas assure orderly and complete coverage of essential matters. We have included a suggested agenda for your lessons learned reviews, assisting in the population of the lessons learned database, and used in throughout the program. A final lessons learned session is mentioned in section 7.1.3.1 in the *Standard for Program Management*—Third Edition (2013).

Lessons Learned Review Agenda Instructions

Lessons learned are collected throughout the program for use on future program activities and in other programs and projects under way in the organization. They are documented in the lessons learned data base and should be accessible to members of the program management team and others given appropriate access rights. Ideally, they should be part of a knowledge management system in the organization. These reviews should be conducted periodically as programs tend to be long, and many team members and other key stakeholders will leave and join the program at different phases of the program's life cycle.

Typical participants in a lessons learned review include the program manager, members of the program management team, a knowledge management specialist in the organization, the director of the program management office, and the portfolio manager. The program manager should review the stakeholder register, stakeholder, and the stakeholder engagement plan and decide whether or not to invite other stakeholders to the meeting.

Prior to the meeting, participants should receive copies of the lessons learned register.

An agenda for these meetings is as follows:

PARTICIPANTS (NAMES/ORGANIZATION)

Date: _____

Time: _____

Place: _____

Program overview: The program manager presents a brief high-level overview of the program as to why the program was undertaken and where it stands in the program management life cycle. After the first meeting, he or she should provide a short summary of the last meeting held and should ensure each participant has the latest updates to the lessons learned data base.

Discussion of the types of lessons learned: The program manager should lead a discussion concerning the types of lessons learned during the meeting using the classification system defined in the knowledge management plan. The purpose is to see whether the classification system is helpful or whether categories should be added or deleted.

Use of lessons learned: The program manager next leads a discussion as to the usefulness of lessons learned in terms of exploiting possible opportunities. The discussion also concentrates on responses to any negative items so preventive action can be taken so they do not occur in the future.

Risk management processes: The program manager also asks the group about the usefulness of these reviews in terms of the various risk management processes where lesson learned reviews often are used.

Suggestions and recommendations: The program manager asks the group for any suggestions or recommendations concerning the lessons learned process so it is more useful to the program and other initiatives in the organization. These suggestions are documented in the program's decision log.

Decisions: This section documents the decisions made during this meeting.

Signatures of Participants

Program manager _____

Director of the program management office _____

Portfolio manager _____

Knowledge management specialist _____

Participant 1 _____

Participant 2 _____

Participant N _____

Lessons Learned Data Base

The *Standard for Program Management*—Third Edition (2013) discusses the lessons learned data base in several sections: 8.1.2.4, Lessons Learned Data Base; 8.3.1.6, Program Roadmap and Charter Development; 8.3.7.2, Knowledge Transition; 8.4.4, Program Procurement Closure; and 8.7.1, Program Risk Management Planning. The lessons learned records are referred to as a data base. Here is a suggested format, no matter the storage medium.

Lessons Learned Data Base Instructions

Purpose: Lessons learned are collected throughout the program. This data base is used to document lessons learned in the program and may be used to support a best practices library or a knowledge repository.

Lessons learned include both weaknesses to correct and be aware of for future programs and projects, and opportunities to be pursued in this program and in other programs and projects in the organization. The program management team should review them regularly, and the program manager should discuss lessons learned with the Governance Board at program review meetings for applicability to other programs and projects under way in the organization. At the end of the program, the lessons learned data base should be part of the program's archives. It should also become part of the organization's knowledge management repository, as these lessons learned are part of the organization's assets and are a compilation of knowledge gained for use on future programs and projects.

The lessons learned data base is designed as a table, and a description of its contents follows.

1. **Lesson learned number:** Assign an identification number to the lesson learned.
2. **PWBS number:** Link the lesson learned to the corresponding program package in the program work breakdown structure.
3. **Description:** Provide a description of the lesson learned. Set up a classification system such as:
    a. Technical
    b. Managerial
    c. Policy
    d. Procedure
    e. Process
        i. Program management
        ii. Stakeholder engagement
        iii. Governance Board
        iv. Benefits delivery

      v. Audits
      vi. Reviews
      vii. Project management
   f. Cause of variance from a plan
   g. Corrective action used and its impact
   h. Effectiveness of selected risk response
      i. Lack of a need to implement a fallback plan
      ii. Lack of subsidiary risks as a result of implementing the response
      iii. Contingency plan implemented and its impact
   i. Issue resolution
   j. Benefit realized
   k. Best practice
   l. Business skills
  m. Unforeseen risk
   n. Product improvement
   o. Service improvement
   p. Evaluation criteria
   q. Qualified seller
   r. Contract terms and conditions
   s. Reason for contract termination

4. **Supporting details:** List supporting details as appropriate.
5. **Key words:** Provide some key words for use as metadata tags for future searching for information that may be contained as part of each lesson learned, especially when it becomes part of the organization's knowledge management repository and the program's archives.
6. **Submitted by:** List the person who submitted the lesson learned and provide contact information. Note that the lesson learned could be submitted as part of a review by the Governance Board or as a recommendation in a program audit or a quality assurance audit.
7. **Date:** State the date the lesson learned was added to the data base.
8. **Program life cycle phase:** List the phase in the program life cycle that involves the lesson learned. The data base should be updated as components end and at the end of the program.
9. **Affected program processes:** List the program management plans and processes affected by the lesson learned.
10. **Program documents affected:** State the program documents affected by the lesson learned and whether any require updates. If updates are needed, state the program management team member who is responsible and the date the documents are to be updated.
11. **Use in the program:** State whether the lesson learned was used later in the program if applicable.
12. **Date submitted to the knowledge management repository or archives:** If the organization has a knowledge management repository, state the

date this lesson learned was submitted to it so it can be used by other programs and projects in the organization. Once submitted, the lesson learned becomes a source of historical information for use by other programs and projects.

13. **Date submitted to archives:** State the date the lessons learned register was submitted to the program's archives.

14. **Notes:** Use this field for any additional notes about the lesson learned.

Decision Log

Significant decisions should be recorded. Decisions and the "whys" of those decisions are sometimes forgotten, often leading to disastrous decision reversals, neglect of important reversals, or do-loop iterations of the same "new" decision. In some cases, your customer may force you to implement a poor decision, and this log is your chance to document that fact and your reservations.

While a governance decision register is noted in the *Standard for Program Management*—Third Edition, other decisions also will be made during the program by people other than the members of the Governance Board; this log then is used for those decisions. Please refer to Chapter 6 for a template for the governance decision register.

Decision Log Instructions

Purpose: The decision log is used throughout the program by the program manager to document the major program decisions. In some organizations, it may be called a memorandum for record. The program manager should decide the level of detail to include in this log. The log is especially useful on large programs in which information about a particular decision made early in the program requires later review or where decisions are forced upon the program managers, for example, by the customer. The log also is helpful if someone else is assigned later as the program manager. The log should be part of the program's final records, as it can serve as an excellent source of lessons learned for future programs.

This decision log is designed as a table, and a description of its contents follows.

1. **Decision number:** Provide a number for the decision for tracking purposes.
2. **PWBS number:** Link the decision to a PWBS program package.
3. **Decision description:** Describe the actual decision and what is to be done. Provide sufficient detail for future use.
4. **Background information:** Provide background information as to why the decision was needed. Examples include an escalated risk by a component manager to the program manager, an issue to be resolved, a customer request for additional work to be done, lack of realization of

benefits as described in the benefits realization plan, a capacity problem, or a missed deliverable.

5. **Decision maker:** List the decision maker and his or her contact information.
6. **Implementation date:** State the date the decision is to be implemented.
7. **Assigned to:** State the person who is responsible for implementing the decision and his or her contact information.
8. **Actual implementation date:** State the actual date the decision was implemented. Provide information if there is a variance between the actual date and the planned date.
9. **Notes:** Use this field for any additional notes about the decision.

Program Issues Register

Issues of various types are common on programs and are discussed throughout the *Standard for Program Management*—Third Edition (2013). An issue log is mentioned in Section 5.3, Stakeholder Engagement, and an escalation process for issues to be handled by the Governance Board is described in Chapter 6 of this book. This register, though, provides a format for recording and reporting on the variety of program issues that may occur throughout the program's life cycle. Further, in the *Standard for Program Management*—Third Edition (2013), Appendix X.4.2, Successful Program Managers, notes the importance of the program manager's ability to manage issues throughout the program.

Program Issues Register Instructions

Purpose: The program issues register is used to track issues on programs, including those to be resolved by members of the Governance Board. Issues involve a variety of areas—legislation, customer requests, benefit realization, schedule, finances, scope, stakeholder concerns, and unresolved risks to list a few.

The program issues register, also known as an issue log, identifies each issue and then shows how it was resolved, and the people who were involved in the process. Issues may be identified by any program stakeholder. Issues at the program level may be ones that are escalated to the program manager for resolution by project managers. At the program level, the issues register shows whether an issue identified by a project manager affects other projects or non-project work in the program. Regardless of the person who identified the issue, the issues register also shows whether the issue impacts other programs, projects, or work under way in the organization. Note that the issues register does not replace the need for a risk register.

Certain issues can be handled by the program manager; others require involvement by other stakeholders or may need resolution by the Governance Board. Often, the decisions regarding program issues lead to the need to revise other planning documents or to take corrective actions. If this is the case, a change request should be used. Issue reviews should be conducted on a periodic basis, and the issues register is a key document when these reviews are held to show the status of identified issues and their resolution. The program issues register is designed as a table, and a description of its contents follows.

1. **Issue identification number:** Assign an identification number to the issue.
2. **PWBS number:** Link the issue to the corresponding program package in the program work breakdown structure.
3. **Issue description:** Describe the issue and why it is significant to the program.
4. **Identified by:** List the person who identified the issue and provide contact information.
5. **Date:** State the date the issue was identified and added to the register.
6. **Issue type:** Describe the type of issue. Examples include:
   a. Internal: Resources, schedule, scope, cost, quality, contracts
   b. External: Environmental, political, ethical
7. **Root cause:** Describe the root cause of the issue in order to determine needed preventive action for the future and corrective action for this issue.
8. **Issue impact:** State the impact of the issue to the program. Examples include:
   a. Impacts other projects in the program
   b. Impacts other projects or programs under way in the organization
   c. Impacts the organization's strategic plan
   d. Impacts areas outside of the organization
9. **Issue owner:** Identify a member of the program management team to be responsible for the issue and to track it until it is resolved and closed. Provide contact information for the owner.
10. **Proposed resolution:** State the proposed resolution of the issue, such as the corrective action that is required. Recognize that the issue owner may need to work with the person who identified the issue and other stakeholders to determine an appropriate way to resolve the issue so it has minimal impacts to the program and other areas.
11. **Decision:** State whether the proposed resolution was approved, deferred, or rejected.
12. **Resolved by:** State who resolved the issue. Examples include:
    a. Issue owner
    b. Program manager
    c. Governance Board

    Note whether the issue was resolved as proposed or whether a different approach was approved; if the latter, state the approach to resolve the issue. Recognize that a change request will need to be used to implement the resolution.
13. **Date:** State the date the issue was resolved.
14. **Subsequent impacts:** State any subsequent impacts as a result of resolving the issue. Examples include the need to modify the scope of the program,

change processes and procedures used by the program, add resources, modify plans, modify program benefits, or change stakeholder expectations.

15. **Date closed:** State the date the resolution to the issue was implemented to close the issue.

16. **Notes:** Use this field for any additional notes about the issue.

Component Charter

A component charter is recommended since programs contain more than just projects.

Component Charter Instructions

Purpose: A brief introductory statement defining the purpose of the charter, such as:

The component charter states the purpose of the component and shows how the component fits into the overall program. It also states the component manager's authority and responsibility and formally authorizes the component.

Components may be projects, non-project work, or a set of sub-programs, projects and sub-projects. Each component requires a charter before it begins work. It is prepared once the component is officially initiated. This charter then guides the work of the component and supports the overall program charter. Although it is prepared when the component is approved, it may need updates as work progresses during the life cycle, when other components are added to the program or are completed or terminated.

If the component was under way before the program began and then became part of the program, its existing charter should be reviewed to see whether changes are required to support the program's vision, objectives, and benefits. During the program, status reports on the progress of each component are prepared.

Component objectives: Each component will have certain objectives. This section describes how the component's objectives support the overall program objectives.

Component expected benefits: Components provide benefits, and as they are managed in a program structure, the incremental benefits from all of the components typically are greater than if the component was managed individually as a project or as non-project work. This section describes the benefits to be realized by the component and how the component's benefits fit within the program's benefits realization plan.

Component deliverables: This section lists specific deliverables in terms of products, services, and results to be provided by the component. The component requirements documents are used to develop the PWBS. Scope statements for each component are considered as procurements are planned along with cost estimates. Component milestones are part of the program's schedule. Any risks associated with completing these deliverables should be identified for inclusion in the program's risk identification process, with component risk responses considered in the program's risk responses that may be used.

Component stakeholders: Since stakeholder engagement is essential for effective program management, this section lists the key stakeholders who may be involved in, have an interest in, or have an influence concerning the specific component. Component stakeholder engagement guidelines are an input to the manage component interfaces process, and they are developed along with the program's stakeholder engagement plan.

Component constraints: Constraints are factors that limit the options of the project team. This section lists constraints that may affect the component.

Related program assumptions: Assumptions are factors that the project team considers to be true, real, or certain in planning the program. This section lists the assumptions that may affect the component.

Component manager's authority: The charter should name the component manager and describe his or her authority and responsibilities, especially in terms of being able to apply resources to the component.

Approvals: This section lists the approvals of the component charter by the component manager, program manager, program sponsor, members of the Governance Board, and other stakeholders as required.

Component Charter Template

<Insert Program Name>
Component Charter

Program name:	
Component manager:	Email address here as a hyperlink
Program manager:	
Program sponsor:	
Actual start date:	
Approved end date:	
Program number:	
Revision history:	
Business unit:	

A. PURPOSE

A brief introductory statement defining the purpose of the component charter, such as:

> The component charter states the purpose of the component and shows how the component fits into the overall program. It also states the component manager's authority and responsibility and formally authorizes the component.

_____

_____

_____

B. COMPONENT OBJECTIVES

This section describes how the component's objectives support the overall program objectives.

_____

_____

_____

C. COMPONENT EXPECTED BENEFITS

This section describes the benefits to be realized by the component and states how the component's benefits fit within the program's benefits realization plan.

_____

_____

_____

D. COMPONENT DELIVERABLES

This section lists specific deliverables in terms of products, services, and results to be provided by the component.

_____

_____

_____

E. COMPONENT STAKEHOLDERS

This section lists the key stakeholders who may be involved in, have an interest in, or have influence concerning the specific component.

_____

_____

_____

F. COMPONENT CONSTRAINTS

This section lists the constraints that may affect the component.

_____

_____

_____

G. RELATED PROGRAM ASSUMPTIONS

This section lists any assumptions from the component that may affect the program.

_____

_____

_____

H. COMPONENT MANAGER'S AUTHORITY

This section names the component manager and describes his or her authority and responsibilities, especially in terms of being able to apply resources to the component.

_____

_____

_____

I. APPROVALS

This section lists the approvals of the component charter by the component manager, program manager, program sponsor, program management office director, members of the Governance Board, and other stakeholders as required.

SIGNATURES AND DATE APPROVAL OBTAINED

Component manager _____

Program manager _____

Program sponsor _____

Program management office director _____

Governance Board chairperson _____

Governance Board member 1 _____

Governance Board member 2 _____

Governance Board member N _____

Stakeholder 1 _____

Stakeholder 2 _____

Stakeholder N _____

Quality Assurance

Quality planning, assurance, and control are continual on programs as explained in the *Standard for Program Management*—Third Edition (2013). At the program level, the emphasis on quality assurance is on cross-program, inter-project relationships. The program manager and his or her team, often working in conjunction with the Quality Department, focuses on a project's quality specifications that may impact another project or the entire program. At the program level, quality assurance emphasizes analysis of the quality control results, see Section 7D, of the various projects and non-project work to foster overall program quality.

Quality Assurance Audits

Quality assurance audits are recommended in section 8.5.2 in the *Standard for Program Management*—Third Edition (2013). While the Governance chapter in this book (Chapter 6) discusses the Quality Management Plan, an Audit Plan, and a format for an Audit Report for overall program audits (6.6.4), a format for a quality assurance audit is presented in this chapter. These quality assurance audits should be included in the Audit Plan, and if they are known in advance, they should be in the program's master schedule. The program manager should convey to his or her team the necessity for these audits and promote their use in a positive way. The recommendations from the quality assurance audit may lead to change requests to improve overall program quality management.

Quality Assurance Audit Report Instructions

The Quality Assurance Audit Report includes the following:

Purpose: A brief introductory statement defining the purpose of the Quality Assurance Audit Report such as:

> The Quality Assurance Audit Report presents an objective assessment of the overall program quality to ensure the program is in compliance with quality policies and standards. Its focus is to ensure quality is a continuous process using the audits to promote best practices

and proper updates. They also are useful because over the length of the program, new laws and regulations may lead to new quality standards, which the program must meet. The report's findings and recommendations serve to enhance program effectiveness and foster process improvement taking into account that every program component contributes to overall program quality, and overall program quality standards must be monitored and controlled. The Quality Assurance Audit Report is an objective report of findings and provides recommended preventive or corrective actions to follow.

Background Information

Time Period of the Audit: This section states the time the audit began and when the final report was prepared.

Place: This section states where the audit was conducted.

Audit Participants: This section states the names of the people on the audit team and provides their contact information for possible questions.

Type of Audit: This section states the type of audit that was conducted: internal or external. Note whether the audit was conducted according to a schedule in the program's master schedule based on the Audit Plan or whether it was random. It describes if the quality assurance audit focus is on the entire program or only on certain components. If it is for certain components, the reason these components were selected is stated.

Reason for the Audit: This section describes why the audit was conducted such as to concentrate on whether a new regulation or standard has been implemented, to determine the extent to which the quality management plan is being followed, or to evaluate the usefulness of the quality control techniques to promote adequate benefit realization. The audit also may evaluate the extent that quality management is an ongoing process on the program and the resources allocated to it.

Methodology: This section describes how the audit was conducted. For example, it notes whether documents were reviewed, interviews were held, surveys were used, focus groups were used, and/or other methods.

Findings: This section presents the findings from the audit. The findings should be presented in an objective way as they are the basis for the recommendations from the audit. They are the synthesis of all of the data that were reviewed. Findings should be carefully worded to reflect the audit team's observation, and they should be phrased as constructively presented problem statements. Appropriate background information that is needed to understand the findings should be included. Causes are observations that support the findings, and their identification is helpful in making constructive recommendations. Consequences list the problem results of a finding. Findings should represent issues for the entire program or for the components being audited and should have the broadest possible application.

Here is an example:

> *Finding (problem statement):* Each status report that is prepared omits mention of quality management activities under way on the program.
>
> *Probable cause:* Program participants lack knowledge of their specific roles and responsibilities for quality management.
>
> *Business consequence:* Deliverables from program components may not meet specific standards set by local authorities and cannot be accepted by the customer.
>
> *Recommendation:* Prepare a Quality Management Plan that describes the importance of quality on the program, sets forth a quality policy, states quality specifications that are to be met, describes quality assurance and control requirements, and states the responsibilities of the program team in quality management. Provide an overview of the Quality Management Plan to each team member via a webinar and make it available in the knowledge repository. Review the plan on a periodic basis to ascertain its usefulness and determine whether changes are warranted. Ask the Quality Department to review the Plan to determine if it conforms to the organization's quality policy.

In preparing findings, keep the specific quality assurance audit goals and appraisal scope in mind. Avoid the following:

Moot issues
Findings based on hearsay alone
Broad generalizations

Recommendations: This section states the recommendations based on the audit findings. Use categories if possible to group the recommendations and present them in priority order. If the audit is for compliance purposes, note if some of the recommendations are mandatory.

Appendices: Include appendices such as the following:

Persons Interviewed: List the people who were interviewed during the audit and the date of each interview

Documents Reviewed: List the documents that were reviewed during the audit

Approvals: This section contains the written approval of the Audit Report by the lead Auditor.

Quality Assurance Audit Report Template

<Insert Program Name>
Audit Report

Program name:	
Program manager:	PM's email address here as a hyperlink
Program sponsor:	
Actual start date:	
Approved end date:	
Program number:	
Revision history:	
Business unit:	

A. PURPOSE

A brief introductory statement defining the purpose of the Quality Assurance Audit Report such as:

> The Quality Assurance Audit Report presents an objective assessment of the overall program quality to ensure the program is in compliance with quality policies and standards. The report's findings and recommendations serve to enhance program effectiveness and foster process improvement taking into account that every program component contributes to overall program quality, and overall program quality standards must be monitored and controlled.

B. BACKGROUND INFORMATION

This section provides background information about the audit. It should describe when the audit began and was completed, where the audit was conducted, who participated on the audit team, the type of audit that was conducted such as for the entire program or only certain components, and the reason the audit was conducted.

_____

_____

_____

C. METHODOLOGY

This section describes how the audit was conducted.

_____

_____

_____

D. FINDINGS

This section presents the findings from the audit.

_____

_____

_____

E. RECOMMENDATIONS

This section states the recommendations based on the audit findings, in priority order, with notes if there are any mandatory recommendations if the audit was conducted for compliance purposes.

_____

_____

_____

F. APPENDICES

This section lists appendices such as the name of people who were interviewed, and the documents that were reviewed.

_____

_____

_____

G. APPROVALS

This section contains the approval of the Audit Report by the lead Auditor.

SIGNATURES AND DATE APPROVAL OBTAINED

Lead Auditor _____

Section 7D: Program Benefits Delivery: Monitoring and Controlling the Program

Drive thy business, let not that drive thee.

—Benjamin Franklin

We now turn to another important part of this process—the business of monitoring and controlling. As stated in the *Standard for Program Management*—Third Edition (2013), it is a responsibility of program managers to monitor the progress of the program and its components to ensure the goals, benefits, budget, and schedule are met (p. 7). Monitoring and controlling the program are discussed throughout the Standard in benefits management, stakeholder engagement, and governance as well as in sections 7.1.2.2, Component Oversight and Integration; 8.1.3, Program Performance Reporting; 8.2.6, Program Financial Monitoring and Control; 8.3.5, Program Performance Monitoring and Control; 8.4.3, Program Procurement Administration; 8.5.3, Program Quality Control; 8.6.3, Resource Interdependency Management; 8.7.5, Program Risk Monitoring and Control; 8.8.2, Program Schedule Control; and 8.9.2, Program Scope Control.

A high-level view of the tasks in Controlling the Program from PMI's *Examination Content Outline* (PMI, 2011) is as follows:

■ Analyzing variances and trends to identify corrective actions or opportunities
■ Updating program plans to incorporate corrective actions
■ Managing program level issues (covered in Section 7C)
■ Managing changes according to the change management plan (covered in Section 7C)
■ Conducting impact assessments for changes and to recommend decisions
■ Managing risk by following the risk management plan

Strongly motivated to run our program rather than having it run us, we can put the following items to work.

Impact Analysis

This analysis tool is helpful in the overall process of monitoring and controlling the program as it facilitates the complete analysis and documentation of proposed changes. Impact analysis is noted in the *Standard for Program Management*—Third Edition (2013) in terms of evaluating the urgency and probability of stakeholder issues. Appearing quite useful, however, in other circumstances, we include it here.

Impact Analysis Instructions

The impact analysis includes the following:

Purpose: A brief introductory statement defining the purpose of the impact analysis, such as:

> The impact analysis describes the factors involved in a program change and is used to explain the decision regarding the change request.

> The impact analysis provides information regarding the change decision, and it considers the program benefits associated with each change. It also is important to recognize that changes may be accepted, deferred, modified, or rejected. The impact analysis, therefore, supports the program's change management plan and change log. Decisions are communicated to stakeholders following the program communications management plan.
>
> The impact analysis is used in other aspects of the program and by various program components.

Type of change: This section lists the type of change in order that its decision then can be communicated to the stakeholders who are interested. In completing this part of the template, the program management team should refer to the types of changes listed in the change management plan.

Factors involved in the change: This section lists the factors involved in the change. For example, a change initially may be viewed as one that affects only one component in the program. When the change request form is prepared and it is analyzed, however, this change may be one that affects multiple components, deliverables, and program benefits. It also may affect other programs or projects under way in the organization. Additionally, it may affect some of the key processes and procedures used in program management in the organization.

Effect on common objectives: This section lists the effect on common objectives at the program level because of the impact analysis. It considers both managerial and technical objectives. This section further considers the interdependencies between program components as well as the impact on the overall strategic objectives.

Change decision: Based on the information provided, changes may be approved, deferred until a later date, modified, or rejected. The person or persons responsible for making the decision concerning the change also may request additional information. This section states the change decision.

Approving authority: This section lists the person or persons who made the decision regarding the change. The program manager may make this decision on his or her own, it may be referred to the Governance Board, or it may be referred to a configuration management board or change control board.

Communication of the change decision: Based on the program communications management plan and the stakeholder engagement plan, certain stakeholders will have an interest in the impact analysis. This section describes those stakeholders who should be notified based on the topic of the change.

Change requests: As part of the impact analysis, change requests will need to be issued. Many of these changes, for example, will require updating the program management plan and perhaps the change management plan. They may require an update to the program work breakdown structure, schedule, financial plan, and others.

Management: This section contains the written approval of the impact analysis by the program manager, program sponsor, members of the Governance Board, and any other key stakeholders as appropriate.

Impact Analysis Template

<Insert Program Name>
Impact Analysis

Program name:	
Program sponsor:	
Program manager:	PM's email address here as a hyperlink
Actual start date:	
Approved end date:	
Program number:	
Revision history:	
Business unit:	

A. PURPOSE

A brief introductory statement defining the purpose of the impact analysis, such as:

The impact analysis describes the factors involved in a program change and is used to explain the decision regarding the change request.

_____

_____

_____

B. TYPE OF CHANGE

This section provides a description of the type of change in order that its decision can be communicated to appropriate stakeholders.

_____

_____

_____

C. FACTORS INVOLVED IN THE CHANGE

This section lists the factors involved in the change.

_____

_____

_____

D. EFFECT ON COMMON OBJECTIVES

This section lists the effect on common objectives at the program level because of the impact analysis.

_____

_____

_____

E. TYPE OF CHANGE DECISION

This section states the change decision.

_____

_____

_____

F. APPROVING AUTHORITY

This section lists the person or persons who made the decision regarding the change.

_____

_____

_____

G. COMMUNICATION OF THE CHANGE DECISION

This section describes those stakeholders who should be notified based on the topic of the change.

_____

_____

_____

H. CHANGE REQUESTS

This section describes the need for any change requests based on the impact analysis.

_____

_____

_____

I. APPROVALS

This section contains the approval of the impact analysis by the program manager, program sponsor, members of the Governance Board, and any other key stakeholders.

SIGNATURES AND DATE APPROVAL OBTAINED

Program manager _____

Program sponsor _____

Governance Board member 1 _____

Governance Board member 2 _____

Governance Board member N _____

Stakeholder 1 _____

Stakeholder 2 _____

Stakeholder N _____

Program Performance Report

The *Standard for Program Management*—Third Edition (2013) talks about program performance reporting in section 8.1.3, noting the necessity of consolidating data across the program and making it available as appropriate to stakeholders to provide assurance to them that resources are being used effectively to deliver program benefits. The sponsor or the contract, if the program is being done under contract, may require specific formats for these reports. Customers also may request periodic feedback. Section 8.3.5, in its discussion of program performance monitoring and control, notes the importance of program performance reports and forecasts to assess trends and determine whether preventive or corrective actions are required. It also uses the *status report* to describe the program's progress of its components and toward benefit realization and identifies resource use concerning benefit realization and meeting the program's goals. Often, these reports contain earned value data, work that is complete according to milestones and dates, work yet to complete, and any risks, issues, and changes that are being considered. Since performance and status reports are used interchangeably, we use the term *performance report*.

Program Performance Report Instructions

Purpose: The program performance report provides a high-level status of program accomplishments and is distributed to various stakeholders in accordance with the program communications management plan. Stakeholders use this report for program monitoring and control and for general information as appropriate, given the stakeholder's level of interest and involvement in the program.

The following describes the items recommended to be part of the performance report:

1. **Program name:** List the name of the program.
2. **Prepared by:** List the name of the person who prepared the report and his or her contact information.
3. **Date:** List the date the report was prepared.
4. **Program budget:** Provide information on the total program budget.
5. **Funds expended to date:** State the amount of funds that have been spent as of the date of the report.
6. **Business case:** Use this section for summary information about the program's business case.
   a. **Date of initial business case:** State the date the initial business case was prepared.
   b. **Changes needed and date:** Describe whether or not any changes were required to reflect changing conditions.
   c. **Assigned to:** List the person who is responsible for updating the business case and the date it is due.
7. **Benefit status:** Use this section for a summary of the status of the expected benefits to be realized by the program.
   a. **Benefit:** List the benefit and any corresponding number used for tracking purposes.
   b. **Description:** Provide a brief description of the benefit.
   c. **Planned realization date:** State the date the benefit is expected to be realized.
   d. **Actual realization date:** State the actual date the benefit was realized.
   e. **Status:** Provide information as to the status. Consider a format such as green for on track, yellow for may not be realized as planned, and red for not expected to be realized.
   f. **Notes:** Use this field for any additional notes or comments about the benefit.
8. Deliverable status: Use this section for a summary of the status of the program's deliverables.
   a. **Deliverable:** List the deliverable and any corresponding number used for tracking purposes.
   b. **Description:** Provide a description of the deliverable.
   c. **Planned date:** State the date the deliverable is expected to be complete.
   d. **Actual date:** State the actual date the deliverable was completed.
   e. **Status:** Provide information as to the status. Consider a format such as green for on track, yellow for may not be completed as planned, and red for not expected to be completed.
   f. **Notes:** Use this field for any notes or comments about the deliverable.

9. **Schedule status:** Use this section for a summary of the status of the program's schedule.
    a. **Milestone and date:** List the milestone and expected date.
    b. **Date completed:** List the date the milestone was finished.
    c. **Schedule revision:** Provide information as to whether a revision to the program's schedule is required.
    d. **Notes:** Use this field for any notes or comments about the milestone.
10. **Change request status:** Use this section for a summary of the program's change requests.
    a. **Description of the change request:** Provide a brief description of the change request. Include a number if used for tracking purposes.
    b. **Submitted date:** State the date the change request was submitted.
    c. **Status and date:** List whether the change request was approved, deferred, or rejected and the date. Also note whether any additional information was required.
    d. **Implementation date:** State the date the change request was implemented, if appropriate, and who was assigned to implement the change.
    e. **Notes:** Use this field for any notes or comments about the change request.
11. **Issue resolution status:** Use this section for a summary of program issues.
    a. **Issue:** Provide a brief description of the issue. Include a number if used for tracking purposes.
    b. **Assigned to:** State the person who is responsible for the issue and provide contact information.
    c. **Proposed resolution and date:** Describe the proposed resolution and the date the issue is scheduled to be resolved.
    d. **Actual resolution and date:** State the actual resolution and the date the issue was resolved.
    e. **Notes:** Use this field for any notes or comments about the issue.
12. **Risk resolution status:** Use this section for a summary of program risks and the status of their resolution.
    a. **Risk:** Provide a brief description of the risk. Include a number if used for tracking purposes.
    b. **Proposed resolution:** State the proposed resolution for the risk following the risk response plan.
    c. **Assigned to:** State the person who is responsible as the risk owner and provide contact information.
    d. **Actual resolution and date:** State the actual resolution for the risk and the date.
    e. **Notes:** Use this field for any notes or comments about the risk.

13. **Resource issues:** Use this section for any information about project resources.
    a. **Description:** Provide a description of the resource requirement.
    b. **Proposed resolution:** If there is a problem in acquiring the resources, state the proposed resolution.
    c. **Assigned to:** State the person who is assigned to resolve the resource issue and provide contact information.
    d. **Actual resolution and date:** State the actual resolution and date.
    e. **Notes:** Use this field for any notes or comments about the resource issue.
14. **Escalated risks and issues:** Use this section for information about escalated risks and issues to the Governance Board or other stakeholders.
    a. **Description:** Provide a description of the escalated risk or issue.
    b. **Date escalated:** State the date the risk or issue was escalated.
    c. **Escalated to:** List the stakeholders who are to resolve the escalated risk or issue.
    d. **Resolution description and date:** State the actual resolution and the date.
    e. **Notes:** Use this field for any comments about the escalated risk or issue.
15. **Earned value metrics:** Use this section to report earned value metrics, as appropriate.
    a. **Schedule variance (SV):** Use this field for the schedule variance.
    b. **Cost variance (CV):** Use this field for the cost variance.
    c. **Schedule performance index (SPI):** Use this field for the schedule performance index.
    d. **Cost performance index (CPI):** Use this field for the cost performance index.
    e. **Estimate at completion (EAC):** Use this field for the estimate at completion.
    f. **Estimate to complete (ETC):** Use this field for the estimate to complete.
    g. **To complete performance index (TCPI):** Use this field for the to complete performance index.
16. **Accomplishments during the reporting period:** Use this section to list key accomplishments during the reporting period and the date.
17. **Planned accomplishments before the next report:** Use this section to list planned accomplishments before the next report is due and lists the planned date.
18. **Additional comments:** Use this field for any additional comments to include in the report.

Program Performance Report Template

Program Name	Prepared by	Date	Program Budget	Funds Expended to Date
Business Case				
Benefit Status				
Deliverable Status				
Schedule Status				
Change Request Status				
Issue Resolution Status				
Risk Resolution Status				
Resource Issues				
Escalated Risks and Issues				
Earned Value Metrics				

Accomplishments during the Reporting Period	
Accomplishment	*Date*

Planned Accomplishments before the Next Report	
Accomplishment	*Planned Date*

Additional Comments

Risk Audit Report

The objectives and timing of risk audits can be specified in the Audit Plan. This is a suggested format for covering all the relevant points in a risk audit report. Risk audits are a recommended best practice in addition to overall program audits.

Risk Audit Report Instructions

The risk audit report includes the following:

Purpose: A brief introductory statement defining the purpose of the risk audit report, such as:

> The risk audit report presents an objective assessment of the program's risk responses according to the overall risk management process used on the program according to the risk management plan. The report's findings and recommendations serve to enhance risk management effectiveness and ensure that the program is following the risk management plan.

Although program audits are conducted on a regular and ad hoc basis throughout the life cycle, risk audits also are recommended to ensure that risk processes and procedures are effective ones and to see if changes are needed so that risks do not turn into problems that impact overall benefits delivery. Risk audits complement the program audits and also any risk reviews that are held. They should be viewed in a positive manner rather than as necessary for compliance or for pinpointing specific problems and associating them to specific individuals.

These audits are planned and noted in the program's master schedule, although the program manager may feel a need to conduct them on an ad hoc basis. They focus on risks at the program level, as project managers may conduct them at the component level. A recommended best practice is for the audit team to review the risk management plan, the risk register, and any minutes from risk reviews prior to the audit. The risk audit report is an objective report of findings and provides recommended preventive or corrective actions to follow. It also is used to ensure that program risk management and risk response processes are being followed.

The following are recommended contents for the risk audit report.

Background information:

Time period of the audit: This section states the time the audit began and when the final report was prepared.

Place: This section states where the audit was conducted.

Audit participants: This section states the names of the people on the audit team and provides their contact information for possible questions.

Type of audit: This section states the type of audit that was conducted—internal or external. Note whether the audit was conducted according to the program's master schedule, as part of a program quality review, or whether it was random.

Reason for the audit: This section describes why the audit was conducted, such as to concentrate on a risk that occurred that was not part of the risk response plan and affected numerous program components; to evaluate the number of workarounds that were needed to address root causes; to determine whether the risk management planning process is effective or requires change; to determine whether new risk categories may be needed, especially as new components become part of the program and other components end; to assess whether the program stakeholders' tolerance for risks has changed; to determine whether the risk budget is adequate; to see whether the risk monitoring and tracking process is effective; and to determine whether stakeholders have requested any additional information other than that regularly reported on risks.

Program areas affected: This section describes the various areas of the program that were involved in the audit, e.g., the entire program or only certain components of it.

Methodology: This section describes how the audit was conducted. For example, it notes whether documents were reviewed, interviews were held, surveys were used, focus groups were used, or other methods were used.

Findings: This section presents the findings from the audit. The findings should be presented in an objective way, as they are the basis for the recommendations from the audit. They are the synthesis of all of the data that were reviewed. Findings should be carefully worded to reflect the audit team's observation, and they should be phrased as problem statements. Appropriate background information that is needed to understand the findings should be included. Causes are observations that support the findings, and their identification is helpful in making constructive recommendations. Consequences list the problem results of a finding. Findings should represent issues for the entire program, or components, according to the scope of the audit, and should have the broadest possible application.

Here is an example:

> *Finding (problem statement)*: Each risk review involves different participants, and many participants are not empowered to make decisions at the review.
>
> *Probable cause*: Designated risk review attendees have not delegated the authority to others to make decisions when they cannot attend the review.
>
> *Business consequence*: The risk reviews are not considered useful meetings, as decisions are not made as a result of the review.

Recommendation: Prepare and follow a risk review agenda. Have the Governance Board appoint members to attend each review that is held, and if the designated attendee is unable to attend a review, ensure that his or her replacement is empowered to make decisions at the meeting.

In preparing findings, keep the specific risk audit goals and appraisal scope in mind. Avoid the following:

Moot issues
Findings based on hearsay alone
Broad generalizations

Recommendations: This section states the recommendations based on the audit findings. Use categories if possible to group the recommendations and present them in priority order. If the audit is for compliance purposes, note whether some of the recommendations are mandatory.

Appendices: Include appendices such as the following:

Persons interviewed: List the people who were interviewed during the audit and the date of each interview.

Documents reviewed: List the documents that were reviewed during the audit.

Approvals: This section contains the written approval of the risk audit report by the lead auditor.

Risk Audit Report Template

<Insert Program Name>
Risk Audit Report

Program name:	
Program manager:	PM's email address here as a hyperlink
Program sponsor:	
Actual start date:	
Approved end date:	
Program number:	
Revision history:	
Business unit:	

A. PURPOSE

A brief introductory statement defining the purpose of the risk audit report, such as:

The risk audit report presents an objective assessment of the program's risk responses according to the risk management plan and the overall risk management process used on the program. The report's findings and recommendations serve to enhance risk management effectiveness and ensure that the program is following the risk management plan.

_____

_____

_____

B. BACKGROUND INFORMATION

This section provides background information about the risk audit. It should describe when the audit began and was completed, where the audit was conducted, who participated on the audit team, the type of audit that was conducted, the reason the audit was conducted, and the program areas involved in the audit.

_____

_____

_____

C. METHODOLOGY

This section describes how the audit was conducted.

_____

_____

_____

D. FINDINGS

This section presents the findings from the audit.

_____

_____

_____

E. RECOMMENDATIONS

This section states the recommendations based on the audit findings, in priority order, with notes on whether there are any mandatory recommendations if the audit was conducted for compliance purposes.

_____

_____

_____

F. APPENDICES

This section lists appendices such as the names of people who were interviewed and the documents that were reviewed.

_____

_____

_____

G. APPROVALS

This section contains the approval of the audit report by the lead auditor.

SIGNATURE AND DATE APPROVAL OBTAINED

Lead auditor   _____

Program Quality Checklist

Although quality matters, such as checklists are specified in component plans, the program also must conduct similar activities to evaluate its products, services, and results deliverables, management results, and cost and schedule performance. The objectives and timing of checklist usage can be specified throughout the duration of the program in the program quality management plan. Here we present a suggested format for covering the relevant points in such a checklist. This checklist can serve to create a quality review meeting agenda or serve as one input to development of a customer satisfaction survey. Involving stakeholders such as customers, sponsors, and benefits benefactors in the checklist completion are an essential best practice.

Program Quality Checklist Instructions

Purpose: A brief introductory statement defining the purpose of the program quality checklist, such as:

> This checklist can be used for agenda development for a program review meeting or serve as an input to a customer, sponsor, or end user satisfaction survey instrument. The resulting information provides essential information of quality results.
>
> The specific areas and the listed item types are those described in the *Standard for Program Management*—Third Edition (2013). The checklist verification questions should be phrased for Yes/No answers, with elaboration in the comments section. For example, "Was deliverable X delivered on time?" Note that multiple questions can be included for a single deliverable, such as "Was deliverable X of suitable quality?"

The following are recommended contents for the program quality checklist.

Area	Item	Verification Question	OK? Y/N	Comments
Products, Services, and Results Deliverables	Deliverable 1			
	Deliverable 1			
	Service 1			
Management Results	Result 1			
	Result 2			
	Result 3			
Performance	Cost			
	Schedule			
Additional Area				
Additional Area				

Section 7E: Program Closure: Closing the Program

> While we often obsess over planning to start and run a program, we so often completely ignore its closure, and pay dearly for it in ongoing costs and lost lessons learned waiting to be learned again and again.
>
> **—Anonymous**

Like Primo in the movie *Analyze This,* if they hit us with "that closure thing," we need to be ready.

In the exhaustion after the heat of battle and with pressures to move on to the next frontier, closure too often gets slighted. Hundreds or thousands of dormant yet unclosed programs may lie around forever. In a vicious cycle, precious few lessons learned may be available to assist in starting up the next new frontier. Worse, expenses may continue to accrue because of unclosed contract matters.

The *Standard for Program Management*—Third Edition (2013) discusses this phase in its life cycle consisting of program transition and program closeout in section 7.1.3. Also in this Standard, section 8.2.7 discusses Program Financial Closure, 8.3.6 discusses Benefits Transition and Benefits Sustainment [in this book in Chapter 4]; 8.3.7 discusses Program Closure; and 8.4.4 discusses Program Procurement Closure.

A high-level description of the tasks in Closing the Program from PMI's *Examination Content Outline* (2011) follows:

- Completing a program performance analysis report
- Obtaining stakeholder approval for program closure
- Executing the transition and closeout of all program and component plans
- Conducting a post-review meeting
- Reporting lessons learned and best practices

Contract Closure Procedure

Focusing of course on program-level contracts, our format for this procedure gives form to the contract closure considerations mentioned in section 8.4.4, Program Procurement Closure in the *Standard for Program Management*—Third Edition (2013) in order that each contract is formally closed, payments are made to contractors, and no issues are outstanding.

Contract Closure Procedure Instructions

Purpose: Programs typically will use a variety of contractors or suppliers. This document describes a procedure to follow to close contracts that have been awarded. The purpose of this procedure is to close the contract according to the contract's terms and conditions and also according to the overall program management methodology.

This procedure should be reviewed periodically and updated as required.

1. As the contract closure process begins, first review the contracts management plan for any items that apply to closing contracts.
2. Next review the program management plan and the procurement management plan to see if they reference any other items not covered in the contracts management plan.
3. Review the specific terms and conditions for closing contracts as outlined in the contract's terms and conditions. Note each one to ensure it is completed.
4. Review all other contract documentation, such as the following:
   a. Supporting schedules
   b. Unapproved contract changes
   c. Technical documentation prepared by the contractor
   d. Performance and status reports
   e. Invoices
   f. Payment records
   g. Warranties
   h. Results of inspections
   i. Results of audits
   j. Results of reviews
5. Conduct a procurement audit to review the entire procurement process for this contractor.
6. See if there are any outstanding issues or change requests and resolve them.
7. Review the program's budget to see if sufficient funds remain that may be needed to pay the contractor's final invoice. If other funds are required, prepare a change request to obtain the needed funding.
8. Conduct a final performance review with the contractor/supplier for lessons learned.
9. Determine whether there are any follow-up activities because of any variations in the deliverables that were completed and provide a list of these activities to the contractor.
10. Agree upon a mutual date in which any follow-up activities will be completed.
11. Document any warranties or guarantees that are to be provided by the contractor.
12. Assess the contract's terms and conditions to determine whether there are any outstanding intellectual property issues that require resolution; if there are outstanding issues, agree upon a plan and a date by which they will be resolved.
13. Pay the contractor's final invoice.
14. Prepare a written document that is signed by the appropriate people in the organization that states the contract is closed, and all deliverables

have been accepted or rejected. Ensure this document is consistent with the contract's terms and conditions. Provide this document to the contractor. Include any required follow-up activities or warranties.

15. Inform stakeholders that the contract has been closed.
16. Review and document lessons learned regarding this contract, the contractor's performance, and the processes and procedures that are used.
17. Determine whether any updates are recommended to various processes that are used in the organization, such as to the:
    a. Procurement management plan
    b. Contracts management plan
    c. Evaluation criteria
    d. Performance and status report templates
    e. Contract terms and conditions
    f. Contract change control system
    g. Issue resolution process
    h. Claims resolution process
    i. Contract payment system
18. Prepare and submit a recommendation to the contracts/procurement department to retain this contractor on the qualified supplier list, drop this contractor from the list, or add the contractor if it is new to the organization and is not on the list.
19. Prepare a complete set of all the relevant documents about this contract, including the following:
    a. Contracts management plan
    b. Contract
    c. Status and performance reports
    d. Minutes from performance reviews
    e. Invoices and payment records
    f. Technical documents prepared
    g. Audit results
    h. Inspection results
    i. Lessons learned document
    j. Final closeout document
20. Index these documents with appropriate metadata tags so they can be easily retrieved for use on other projects in this program, by other programs and projects under way in the organization, or for future programs and projects.

Supplier Performance Review

When finalizing a program contract, care must be taken to ensure that everything the program needs has been satisfactorily obtained or at least resolved as an issue. For future purposes, we also need to understand how well the supplier performed during this engagement. Here we present a format to help ensure a successful review of supplier performance.

Supplier Performance Review Agenda Instructions

As part of the procurement closure process, the member of the program management team who is serving as the contract administrator for each supplier should conduct a final review of the supplier's overall performance. These reviews provide a number of advantages, both for the performing organization and for the supplier.

The program management team discusses overall performance in terms of objectives that were met, benefits that were achieved, and progress according to plans. This review may be the last time to formally discuss any deficiencies with the supplier. The program management team member then summarizes the results of the review and makes recommendations to the organization's procurement or contracts department concerning adding this supplier to the qualified supplier list if it was a new supplier or continuing to keep the supplier on the list if the supplier has worked for the organization in the past.

The supplier can use the review to discuss any problems that occurred that affected overall performance and any opportunities should the supplier be selected to work on a similar type of contract in the future. This review does not substitute for other reviews that may be held as described in the contracts management plan.

An agenda for these reviews is as follows:

PARTICIPANTS (NAMES/ORGANIZATION)

Date: _____

Time: _____

Place: _____

Contract overview: A member of the program management team, ideally the person who prepared the contracts management plan and who worked with the supplier throughout the contract as the program's contract administrator, presents a brief overview of the contract. He or she describes why it was needed, its deliverables, and impact to the program. He or she also describes why the contract now is complete.

Specific performance requirements: The next part of the review discusses the specific performance requirements to be met by the contract. The program management team reviews each deliverable and describes whether it was completed according to plan. If the planned delivery date was missed, the supplier representatives can explain needed preventive actions. The program management team can discuss the impact of a missed delivery date on the overall success of the program, especially if a missed delivery date affected other parts of the program adversely. Lessons learned can be shared by both parties to improve future performance on similar contracts.

Monitoring and control processes used: During this part of the review, the program management team discusses the various monitoring and control processes used on this contract, such as meetings and status reports. The format for the reports and the frequency in which they were submitted can be discussed. The supplier representatives discuss whether they found these meetings and reports to be useful and present any ideas for change. The program management team can discuss whether change requests were required because of the status reports that were submitted or the meetings that were held. Both groups can discuss whether there was an atmosphere of trust between the two parties and ways in which trust could be enhanced in the future.

Change control: Although changes may be needed as a result of the reports and reviews, suppliers often submit change requests. During this part of the review, the change control system can be discussed to see if there are any recommended changes to improve its effectiveness. The supplier representatives can discuss why they needed to submit change requests, if this was done during the program. They can comment on items such as the length of time it took to submit a request and how they were notified whether the request was approved, deferred, or rejected, and also whether they needed to submit additional information to the program management team to analyze the impact of the change request on the program.

Claims process: If there were a number of contract claims, the claims process should be reviewed to see why many claims were submitted. Both groups can discuss the effectiveness of this process to see if any changes would have facilitated the resolution of claims that were submitted.

Invoicing and payment process: The invoicing and payment processes also can be discussed. Both groups can determine whether any changes could have led to a smoother process.

Additional comments: The program management team provides any additional comments about the supplier performance, and the supplier representatives also provide comments as appropriate.

Signatures of Participants

Program manager	_____
Project manager	_____
Program management team member 1	_____
Program management team member 2	_____
Program management team member N	_____
Supplier representative 1	_____
Supplier representative 2	_____
Supplier representative N	_____

Contract Closeout Report

During your program, you inevitably learn lessons about your program procurement processes and about your suppliers. This report structure helps you record those procurement-specific lessons learned. They will be useful in future procurements, and this report can be prepared as noted in section 8.4.4, Program Procurement Closure, in the *Standard for Program Management*—Third Edition (2013).

Contract Closeout Report Instructions

Purpose: Contractors/suppliers prepare procurement performance reports throughout the life of the contract according to the contractual terms and conditions. The program management team also outlines status reporting requirements in the contracts management plan. The team and other stakeholders use these reports for contract monitoring and control, to ensure that any issues that are raised are resolved in a timely way, and for general information as appropriate. The procurement performance reports are used by the program management team to ensure that all issues raised in these reports are resolved in a satisfactory way before the contract is closed, and the team prepares the closeout report.

In the *Standard for Program Management*—Third Edition (2013), a contract closeout report also is an output from program procurement closure to show the results of the supplier performance review and to record any issues as needed in the program issues register. Additionally, any lessons learned are included in this report.

The following describes the recommended items to be part of the Final Contract Closeout Report.

1. **Contract Number:** List the contract number.
2. **Prepared By:** List the name of the person who prepared the report and his or her contact information.
3. **Date:** List the date the report was prepared.
4. **Report Recipients:** State the recipients of the report.
5. **Contract Overview:** Provide a brief summary of the contract and its purpose.
6. **Deliverable Status:** Provide summary information about the various contractual deliverables. Consider the following:
   a. **Deliverable:** List the deliverable and any corresponding number used for tracking purposes.
   b. **Description:** Provide a description of the deliverable.
   c. **Completion Date:** State the actual date the deliverable was completed.
   d. **Notes:** Use this field for any notes or comments about the deliverable.
7. **Schedule Status:** Provide information for lessons learned concerning the schedule for the deliverables. State any schedule revisions that were required and why they were needed.
8. **Budget Status:** Provide information on the total budget for the contract. Note whether any additional funding was needed to complete the contract.
9. **Change Request Status:** Use this section for a summary of any contract change requests.
   a. **Description of the Change Request:** Provide a brief description of the change request.
   b. **Submitted Date:** State the date the change request was submitted.

    c. **Status and Date:** List whether the change request was approved, deferred, or rejected and the date.

    d. **Implementation Date:** State the date the change request was implemented, if appropriate.

    e. **Notes:** Use this field for any notes or comments about the change request.

10. **Issue Resolution Status:** Use this section to summarize how issues were resolved during the contract. Consider including the following:

    a. **Issue:** Provide a brief description of the issue.

    b. **Assigned To:** State the person who was responsible for resolving the issue and provide contact information.

    c. **Actual Resolution and Date:** State the actual resolution and the date the issue was resolved.

    d. **Notes:** Use this field for any notes or comments about the issue.

11. **Risk Resolution Status:** Use this section to summarize how risks were resolved during the contract. Consider including the following:

    a. **Risk:** Provide a brief description of the risk.

    b. **Assigned To:** State the person who was responsible for resolving the risk and provide contact information.

    c. **Actual Resolution and Date:** State the actual resolution for the risk and the date.

    d. **Notes:** Use this field for any notes or comments about the risk.

12. **Resource Issues:** Describe any resource issues that affected the contractual deliverables. Consider including the following:

    a. **Description:** Provide a description of the resource requirement.

    b. **Assigned To:** State the person who resolved the resource issue and provide contact information.

    c. **Actual Resolution Date:** State the actual resolution and date.

    d. **Notes:** Use this field for any notes or comments about the resource issue.

13. **Earned Value Metrics:** Use this section to report final earned value metrics, as appropriate.

    a. **Cost Variance (CV):** Use this field for the Cost Variance.

    b. **Cost Performance Index:** Use this field for the Cost Performance Index.

    c. **Estimate at Complete (EAC):** Use this field for the Estimate at Completion.

    d. **Estimate to Complete (ETC):** Use this field for the Estimate to Complete.

14. **Results of the Contractor Final Performance Review:** Provide a summary of the final contractor/supplier performance review. Include the date of the review and the people who were present with their contact information.

15. **Procurement Process Recommendations:** Use this section for recommendations regarding updates to the program procurement process. Consider items such as the following:

a. Make-or-buy analysis
b. Procurement management plan
c. Contracts management plan
d. Criteria for inclusion on the qualified seller list
e. Evaluation criteria for source selection
f. Contractual terms and conditions
g. Contractor performance report format, information included, and frequency
h. Contractor performance reviews agenda information, items covered, and frequency
i. Interaction between the program management team and the contractor staff
j. Other lessons learned specific to program procurement management
16. **Additional Comments:** Use this field for any additional comments to include in the report.

Final Contract Closeout Report Template

<Insert Program Name>
Final Contract Closeout Report

Contract Number	Prepared by	Date	Report Recipients
Contract Overview			
Deliverable Status			
Schedule Status			
Budget Status			
Change Request Status			
Issue Resolution Status			
Risk Resolution Status			
Resource Issues			
Earned Value Metrics			
Results of the Final Performance Review			
Procurement Process Recommendations			
Additional Comments			

Program Final Report

At the end of any program, a final report should be prepared to document the work of the program and to be part of the program's archives. The need for a final report to document critical program information is described in section 8.3.7.1, Final Reports, in the *Standard for Program Management*—Third Edition (2013).

Program Final Report Instructions

The program final report includes the following:

Purpose: A brief introductory statement defining the purpose of the program final report, such as:

> The program final report contains the key information about the program, which will be useful to other programs and projects under way in the organization and for future programs and projects.

> In the Project Management Institute's *Standard for Program Management*—Third Edition (2013), the program final report is also used to provide data for corporate governance.

Although it is finalized during the closing process, it should be prepared throughout the program so key information is not forgotten. One best practice to follow is to work on this report at the end of each of the lessons learned reviews and governance reviews that are held. Information from the governance decision register, program decision log, risk register, program issues register, and lessons learned register is reviewed as the report is prepared. Audit, benefit, status, and performance reports also should be reviewed. Reasons for terminating a contract or a component early should be analyzed. Often, program managers will hold a knowledge transition meeting with team members before the program is officially closed; if this type of information is held, it may provide other data to include in the final report.

Suggested items to include in the report are the following:

Program strategic objectives: The program's strategic objectives were described initially in the business case and program mandate. This section states whether or not these objectives were met. If the objectives changed or were not met, this section describes the circumstances for future use by other programs and projects.

Program benefits delivered: Using the benefits realization plan and the benefits reports, prepared throughout the program, this section states whether the intended benefits were met. If they were not, or if other benefits were added as the program ensued, it describes the specific circumstances involved.

Program participants: Since programs often are long, this section lists the people who performed key roles in the program, and the times they held the various positions: program manager, program sponsor, component managers, members of the program management office, as applicable, and members of the Governance Board.

Stakeholder management: This section describes the stakeholders who were involved in, had an interest in, or influenced the program. It uses the stakeholder inventory, stakeholder engagement plan, stakeholder register, and communications log/stakeholder log to summarize how stakeholders were identified, how involved they were in the program, and their importance to the program. It also describes the approaches used by the program management team to work with the key stakeholders in the program.

Summary of program metrics: Throughout the program, a variety of metrics were collected to see whether preventive or corrective actions were required. This section provides a summary of these metrics.

Earned value: Recognizing that the schedule performance index will be 1, this section lists the other EV metrics, such as cost performance index, budget at completion, estimate to complete, estimate at completion, earned schedule, and total cost performance index.

Financial management: This section lists the various financial management metrics collected in the program, such as return on investment, shareholder value added, net present value, payback period, time to break even, internal rate of return, and benefit-cost ratio.

Resources: This section provides information on the planned versus actual resources used in the program, noting that resources are more than only people using the program resource plan as a reference.

Team management: As team members leave the program, an excellent approach to follow is to survey them on areas for improvement. Other surveys can be used periodically to assess the needed competencies for people to the program, the managerial style used, the use of collaborative leadership on the team, and overall team maturity. This section describes the type of team management metrics that were used and makes recommendations for improvement as appropriate. It also describes whether a team-based reward and recognition system was used and its effectiveness.

Charters: This section discusses the effectiveness of the program, component, and team charters and contains recommendations for their improvement.

Organizational: This section discusses any organizational-type metrics that were used in the program, such as the probability of success of the program deliverables, the strategic importance of the program and whether it changed over the course of the program, and the duration of the program as compared to the urgency of the need for its deliverables and benefits.

Program risks: This section describes program risks that were identified and the appropriateness of the risk response that was selected. It includes both opportunities and threats to the program. This section also includes any risks that were not identified but occurred and affected the program. It states the contingency plans that were used. It uses the risk management plan, the risk register, risk reviews, and audits as references.

Program issues: This section describes the key issues that affected the program, using the program issues register as a reference.

Escalated risks and issues: This section lists those risks and issues that the program manager escalated to the Governance Board for resolution and describes the decisions that were made and approaches that were used.

Baseline changes: This section describes the changes over the life of the program to the scope, schedule, cost, benefit, and quality baselines.

Interfaces: This section summarizes the interfaces in terms of interrelationships within the program, with other projects and programs under way in the organization, and organizational, technical, interpersonal, logistical, and political interfaces that occurred.

Communications: This section describes the communications methods used in the program to keep stakeholders informed. It notes any approaches that were added that were not in the program communications management plan. It uses the communications log as a reference. It describes the appropriateness of the meetings that were held and notes whether changes were needed to the agendas for meetings of the Governance Board for phase gate and program reviews as well as the agendas for the risk management planning process and risk reviews. It lists the people who will receive copies of this final report.

Supplier performance: This section summarizes the performance of the various program suppliers and presents recommendations in terms of updating or changing the qualified supplier list as appropriate. It uses the contracts management plan as a reference.

Program decisions: This section describes the key decisions made during the program's life cycle using the decision log and the governance decision register as references.

Terminations: This section summarizes any terminations that occurred during the program.

    Contract: This section summarizes any contracts that were terminated prematurely, to focus on why they were terminated using the procurement program reports as a reference.

    Component: This section summarizes any components that were terminated prematurely, to focus on why they were terminated using the component termination requests as a reference.

Program acceptance: This section lists the formal acceptance of the program's scope to show that each deliverable was completed satisfactorily by the customers and the program sponsor.

Benefits transitioned: This section describes specific benefits that were transitioned to the customer, users, or to a functional unit in the organization, such as a product support group or customer support group, and describes how they will be sustained. It uses the component transition decisions, the transition management plan, and the benefits sustainment plan as references.

Lessons learned: This section summarizes the lessons learned in the program, focusing on successes, failures, and areas of improvement using the lessons learned register as a reference.

    Program: This section summarizes technical lessons learned. It uses the program requirements document as a reference.

    Program management process: This section summarizes the lessons learned in terms of the processes, procedures, and guidelines that were used in the program. It uses the program roadmap and the Program Management Plan as a reference.

Knowledge repository: To best contribute to the organization's knowledge management system and repository, this section contains the metadata tags to locate information about this program in an easy-to-use way as its archives become part of the organization's knowledge repository.

Approvals: This section contains the written approval of the program final report by the program sponsor, program manager, program management office director, members of the Governance Board, and any other key stakeholders as appropriate.

Program Final Report Template

<Insert Program Name>
Final Report

Program name:	
Program manager:	PM's email address here as a hyperlink
Program sponsor:	
Actual start date:	
Approved end date:	
Actual end date:	
Program number:	
Revision history:	
Business unit:	

A. PURPOSE

A brief introductory statement defining the purpose of the program final report, such as:

> The program final report contains the key information about the program, which will be useful to other programs and projects under way in the organization and for future programs and projects.

_____

_____

_____

B. PROGRAM STRATEGIC OBJECTIVES

This section states whether the initially defined strategic objectives for the program were met. If the objectives changed, it describes the reasons for the changes for future use by other programs and projects.

_____

_____

_____

C. PROGRAM BENEFITS DELIVERED

This section states whether the intended benefits were met. If they were not, or if other benefits were added over the program's life cycle, it describes the specific circumstances involved.

_____

_____

_____

D. PROGRAM PARTICIPANTS

This section lists the people who performed key roles in the program and the times they held the various positions.

_____

_____

_____

E. STAKEHOLDER MANAGEMENT

This section describes the stakeholders who were involved in, had an interest in, or influenced the program. It also describes the approaches used by the program management team to work with the key program stakeholders.

_____

_____

_____

F. SUMMARY OF PROGRAM METRICS

This section summarizes the program metrics that were collected.

_____

_____

_____

G. PROGRAM RISKS

This section describes the risks that were identified and the appropriateness of the risk responses that were selected. It also includes any risks that were not identified but occurred and affected the program. It states contingency plans that were used.

_____

_____

_____

H. PROGRAM ISSUES

This section describes the key issues that affected the program.

_____

_____

_____

I. ESCALATED RISKS AND ISSUES

This section lists those risks and issues the program manager escalated to the Governance Board for resolution, describes the decisions that were made and approaches used.

_____

_____

_____

J. BASELINE CHANGES

This section describes changes over the life of the program to the technical, schedule, cost, benefit, and quality baselines.

_____

_____

_____

K. INTERFACES

This section summarizes the interfaces in terms of interrelationships within the program, with other programs and projects under way in the organization, and the organizational, technical, interpersonal, logistical, and political interfaces that occurred.

_____

_____

_____

L. COMMUNICATIONS

This section describes the communications methods used in the program to keep stakeholders informed, the approaches that were added, the appropriateness of the meetings that were held, the agendas that were used, and lists those stakeholders who will receive copies of the final report.

_____

_____

_____

M. SUPPLIER PERFORMANCE

This section summarizes supplier performance and presents recommendations to update or change the qualified supplier list as appropriate.

_____

_____

_____

N. PROGRAM DECISIONS

This section lists the key decisions made during the program's life cycle.

_____

_____

_____

O. TERMINATIONS

This section describes any contracts and components that were terminated prematurely, and the reasons they were terminated.

_____

_____

_____

P. PROGRAM ACCEPTANCE

This section lists the formal acceptance of the program's scope by the customers and the program sponsor.

_____

_____

_____

Q. BENEFITS TRANSITIONED

This section describes specific benefits transitioned to the customer, users, or to a functional unit in the organization and describes how they will be sustained.

_____

_____

_____

R. LESSONS LEARNED

This section summarizes lessons learned, focusing on successes, failures, and areas of improvement in terms of the program and the program management processes used.

_____

_____

_____

S. KNOWLEDGE REPOSITORY

This section contains the metadata tags to locate information about this program in an easy-to-use way as its archives become part of the organization's knowledge repository.

_____

_____

_____

T. APPROVALS

This section contains the approval of the program final report by the program sponsor, program manager, program management office director, members of the Governance Board, and other key stakeholders as required.

SIGNATURES AND DATE APPROVAL OBTAINED

Program manager _____

Program sponsor _____

Program management office director _____

Governance Board chairperson _____

Governance Board member 1 _____

Governance Board member 2 _____

Governance Board member N _____

Stakeholder 1 _____

Stakeholder 2 _____

Stakeholder N _____

Knowledge Transition

Knowledge management is especially important in program management. As part of program closure in the *Standard for Program Management*—Third Edition (2013), knowledge transition is discussed in section 8.3.7.2. We suggest a knowledge transition report be prepared as another way to evaluate overall program performance.

Knowledge Transition Report Instructions

The knowledge transition report includes the following:

Purpose: A brief introductory statement defining the purpose of the knowledge transition report, such as:

> The knowledge transition report is an assessment of the program's performance and is used so the program manager can share lessons learned with the team and others in the organization.

> In the Project Management Institute's *Standard for Program Management*—Third Edition (2013), a knowledge transition report is noted as a meeting the program manager holds to share lessons learned with the program team. Any additional lessons learned then are included in the program's final report.
> We suggest a separate report that is prepared early in the program and enhanced each time the program manager holds a lessons learned review

session as well as at the end of the program. In preparing it, the program manager should review the knowledge management plan, the component transition decisions, the lessons learned data base, and the results of lessons learned reviews. When this report is complete, it may lead to the need to update some of the organization's existing policies, procedures, and processes in program management. The program manager may wish to prepare this report in a meeting with the program management team to formally discuss lessons learned.

Knowledge management goals and objectives: This section lists the goals and objectives for knowledge management based on the knowledge management plan and assesses whether or not they were met. Recommendations for changes are noted. It notes whether people in the program contributed content that was useful to others. The emphasis is to focus on continuous improvement and learning for future programs and projects in the organization.

Program management plans: This section discusses the various plans that the program management team prepared to evaluate their usefulness. It also reviews whether changes were required to these plans and the improvements as a result of these changes. It evaluates the format of the plans, and the people who were involved in the preparation of each plan for future improvements. It assesses the transition plan in terms of knowledge transfer activities to support ongoing benefits in order that the support organizations have required documentation, training, or materials.

Meetings and reviews: This section discusses the effectiveness of the program management team meetings that were held and the performance reviews that were conducted by the team, by the Governance Board, and by auditors. It notes whether the recommendations from the reviews were implemented. The program's decision log and governance decision register are reviewed.

Monitoring and controlling processes: This section describes the various monitoring and controlling processes the program team used to evaluate overall program performance and that of the various program components. It reviews the risk register, program issues register, metrics that were collected, status reports, and the lessons learned data base.

Team performance: A motivated team is essential to program success. This section reviews team performance to determine whether other approaches could have enhanced overall team motivation. It evaluates the usefulness of the team charter, exit interviews with members of the team when they left the program, and rewards and recognition systems in place for program team members.

Contractors and suppliers: Since most programs use a variety of contractors/ suppliers, this section discusses the performance of the various contractors and suppliers with a focus on the effectiveness of the procurement management plan, the contracts management plan, the procurement performance reports, and the contract closure procedure.

Stakeholder engagement: Effective stakeholder engagement is a key to overall program success and benefits realization. This section assesses the stakeholder engagement plan that was prepared. It reviews the stakeholder register and the communications management plan. It notes changes in the various stakeholders throughout the life of the program and reviews the communications log to see the usefulness of the regular reports that stakeholders received, especially if additional reports were required.

Governance reviews: The importance of an effective Governance Board to overall program success cannot be overstated. This section describes the interaction between the program management team and the members of the Governance Board. It evaluates the usefulness of the meetings of the Governance Board both at stage gates and for overall program performance reviews. It also assesses the governance management plan and reviews the governance decision register.

Customer involvement: Customers are active participants in many programs. This section discusses customer involvement apart from stakeholder engagement to see how the program management team interacted with the customers. It also assesses how benefits were transitioned to the customers and how customer satisfaction was measured.

Approvals: This section contains the written approval of the knowledge transition report by the program sponsor, program manager, program management office director, members of the Governance Board, and any other key stakeholders as appropriate.

Knowledge Transition Report Template

<Insert Program Name>
Knowledge Transition Report

Program name:	
Program manager:	PM's email address here as a hyperlink
Program sponsor:	
Actual start date:	
Approved end date:	
Program number:	
Revision history:	
Business unit:	

A. PURPOSE

A brief introductory statement defining the purpose of the knowledge transition report, such as:

The knowledge transition report is an assessment of the program's performance and is used so the program manager can share lessons learned with the team and others in the organization.

_____

_____

_____

B. KNOWLEDGE MANAGEMENT GOALS AND OBJECTIVES

This section lists the goals and objectives for knowledge management and assesses whether or not they were met.

_____

_____

_____

C. PROGRAM MANAGEMENT PLANS

This section discusses the various plans the program management team prepared and their usefulness.

_____

_____

_____

D. MEETINGS AND REVIEWS

This section discusses the effectiveness of the program management team meetings that were held and the performance reviews that were conducted by the team, the Governance Board, and auditors.

_____

_____

_____

E. MONITORING AND CONTROLLING PROCESSES

This section describes the various monitoring and controlling processes the program management team used to evaluate overall program performance and that of the program components.

_____

_____

_____

F. TEAM PERFORMANCE

This section reviews program team performance to determine whether other approaches could have enhanced team motivation.

_____

_____

_____

G. CONTRACTORS AND SUPPLIERS

This section discusses the performance of the various contractors and suppliers with an emphasis on the effectiveness of the procurement management plan, the contracts management plan, procurement performance reports, and contract closure procedures.

_____

_____

_____

H. STAKEHOLDER ENGAGEMENT

This section assesses the stakeholder engagement plan that was used in the program.

_____

_____

_____

I. GOVERNANCE REVIEWS

This section discusses the interaction between the program management team and the Governance Board.

_____

_____

_____

J. CUSTOMER INVOLVEMENT

This section discusses customer involvement apart from stakeholder engagement to see how effectively the program management team interacted with its customers.

_____

_____

_____

K. APPROVALS

This section contains the approval of the knowledge transition report by the program sponsor, program manager, program management office director, members of the Governance Board, and other key stakeholders as required.

SIGNATURES AND DATE APPROVAL OBTAINED

Program manager   _____

Program sponsor   _____

Program management office director   _____

Governance Board chairperson   _____

Governance Board member 1   _____

Governance Board member 2   _____

Governance Board member N   _____

Stakeholder 1   _____

Stakeholder 2   _____

Stakeholder N   _____

References

1. *Analyze This*, Village Road Show Pictures, 1999.
2. Dr. Seuss, *Oh, the Places You'll Go!* New York: Random House, 1993.
3. Project Management Institute. *A Guide to the Project Management Body of Knowledge (PMBOK® Guide)*—Fifth Edition. Project Management Institute: Newtown Square, PA, 2013.
4. Project Management Institute. *Examination Content Outline*. Project Management Institute: Newtown, Square, PA, 2011.
5. Project Management Institute. *The Standard for Program Management*—Second Edition. Project Management Institute: Newtown Square, PA, 2008.
6. Project Management Institute. *The Standard for Program Management*—Third Edition. Project Management Institute: Newtown Square, PA, 2013.
7. Thiry, M. *Program Management*. Gower Publishing Limited: Surrey, England, 2010.
8. Thiry, M. and Deguire, M. *Program Management as an Emergent Order Phenomenon in Innovations: Project Management Research 2004 (Proceedings of the 3rd PMI Research Conference, London, England)*. Newtown Square, PA: Project Management Institute, 2004.
9. United States Department of Defense, *Defense Acquisition Guidebook*, 2004.
10. United States Department of Defense. *Integrated Master Plan and Integrated Master Schedule Preparation and Use Guide*, 2005.

Index

Printed in the United States
by Baker & Taylor Publisher Services

Printed in the United States
by Baker & Taylor Publisher Services